Seneca East Jr. High School

$5.72

**THE ELECTRON MICROSCOPE**

# THE ELECTRON MICROSCOPE
## A Tool of Discovery

**AARON E. KLEIN**

McGRAW-HILL BOOK COMPANY
*New York / St. Louis / San Francisco / London
Montreal / Panama / Toronto*

**Library of Congress Cataloging in Publication Data**

Klein, Aaron E
  The electron microscope.

  (Major tools of science)
  SUMMARY: Traces the development of the electron microscope, how it works, and the major discoveries that have been made with it.
  1. Electron microscope—Juvenile literature.
  [1. Electron microscope. 2. Microscope and microscopy] I. Title.
  QH212.E4K55      502.8      73-17756
  ISBN 0-07-035029-9 (lib. bdg.)

Copyright © 1974 by Aaron E. Klein. All Rights Reserved. Printed in the United States of America. No part of this publication may be reproduced, stored in a retrieval system, or transmitted, in any form or by any means, electronic, mechanical, photocopying, recording, or otherwise, without the prior written permission of the publisher.
12345678    BPBP   78987654

# CONTENTS

1. TO SEE WHAT ESCAPES THE EYE .................. 1
2. A NEW WORLD SEEN AND UNSEEN ................ 9
3. FROM LIGHT MICROSCOPE TO ELECTRON MICROSCOPE ........................................ 15
4. TRANSMISSION ELECTRON MICROSCOPES .......... 29
5. TECHNIQUES ..................................... 41
6. OTHER TYPES OF ELECTRON MICROSCOPES ........ 71
7. KNOWLEDGE OBTAINED WITH ELECTRON MICROSCOPES ..................................... 77

BIBLIOGRAPHY ..................................... 83

INDEX ............................................. 84

Pictures appearing on pages 42, 44, 46, 49, 64, 65, 69, 70
*Courtesy of the ETEC Corporation.*

Pictures appearing on pages 2, 39
*Courtesy of the American Optical Corporation.*

Pictures appearing on pages 50, 51, 52, 53, 66, 69
*Courtesy of American Metals Research (AMR).*

Pictures appearing on pages 20, 22
*Courtesy of Dr. Robert King, Massachusetts Institute of Technology.*

Picture appearing on page 35
*Courtesy of the Kinney Vacuum Company.*

Pictures appearing on pages 54 through 63
*Courtesy of Leonard Phillips.*

Pictures appearing on pages 11, 16, 17
*Courtesy of the Siemens Aktiengesellschaft Company.*

Picture appearing on page 19
*Courtesy of the Parke-Davis Company.*

Pictures appearing on pages 19 bottom, 26, 27
*Courtesy of Japan Electron Optics Limited (JEOL).*

Pictures appearing on pages 31, 35
*Courtesy of McGraw-Hill Book Company.*

Picture appearing on page 74 top
*Courtesy of Dr. Erwin Müller, Pennsylvania State University.*

Picture appearing on page 74 bottom
*Courtesy of Dr. Albert V. Crewe, University of Chicago.*

Picture appearing on page 78 top
*Courtesy of Inger Angerer Klein, Yale University.*

Picture appearing on page 78 bottom
*Courtesy of Dr. R. Wyckoff, National Institute of Health.*

Picture appearing on page 80 top
*Courtesy of Dr. George E. Palade and Dr. Keith Porter from* THE CELL, *Upjohn Company.*

Picture appearing on pages 80 bottom, 81
*Courtesy of Dr. George E. Palade, Rockefeller Institute.*

*To*
*Mickey and Steven*

# Acknowledgments

The picture essay (pages 54 to 63) on preparation techniques was made possible through the cooperation of Dr. Robert King and Elaine V. Lenk of the Massachusetts Institute of Technology. The author also wishes to acknowledge the assistance of the following manufacturers of electron microscopes and electron microscope accessories: American Metals Research Company, American Optical Company, ETEC Corporation, Japan Electron Optics Laboratories, Kinney Vacuum Company, and Siemens Aktiengesellschaft Company.

# 1
# To See What Escapes the Eye

As hammer, saw, and level are tools for the carpenter, so are scientific instruments tools for the scientist. Like hammer and saw, some scientific instruments enable scientists to do things that cannot be done by hands alone. And as the level enables the carpenter to observe and measure what the unaided senses cannot, so too do many scientific instruments.

Of course, the tools of the scientist are, in general, more complex and sophisticated than those of a carpenter. But as complex as the tools of the scientist are, they are intended to do two basic jobs—to extend the senses and to effect changes that cannot be brought about by hands alone. Among the latter are devices that range from simple devices for cutting glass tubing to cyclotrons and accelerators that smash atoms. Instruments that extend the senses can be as simple as a magnifying glass or as complicated as an electron microscope or as marvelously intricate as instruments that can detect the presence of a single particle of an atom that exists for less than a thousandth of a second.

Microscopes were among the first instruments developed by scientists to extend the senses and powers of observation. They have been used since the seventeenth century. Among all the modern, sophisticated scientific instruments the microscope in all its forms is still one of the most useful instruments available to scientists. Almost every kind of scientist uses some kind of

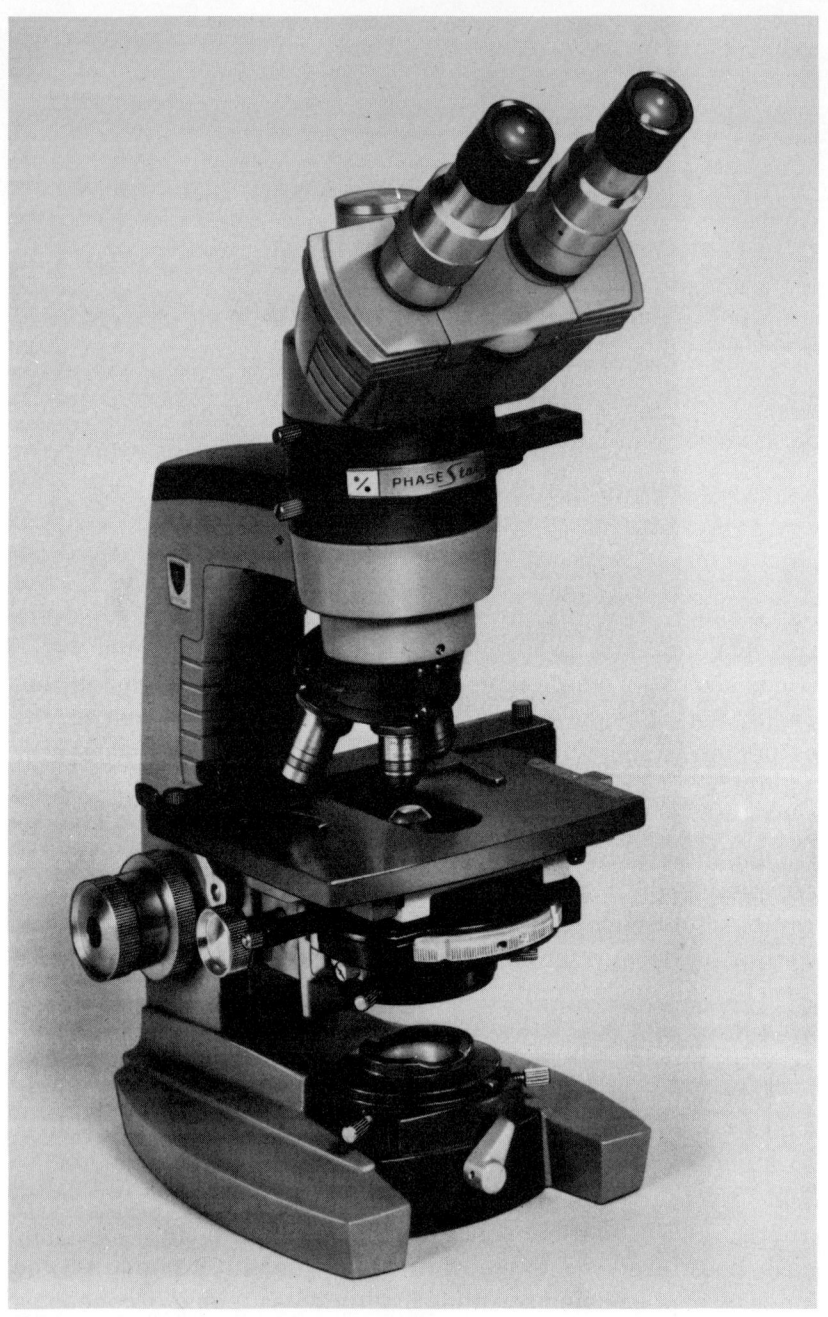

A typical modern light microscope equipped with binocular eyepiece and phase contrast.

microscope in his work. Biologists examine minute living things and the cells and tissues that make up living things. Metallurgists use microscopes to examine the fine structure of metals. Scientists employed in all industries use microscopes to find out more about the materials that go into the products we use.

The light microscope is probably the most familiar of all laboratory instruments. Almost every schoolchild has used one in the course of his education, and the light microscope has come to be the symbol of the scientist and his work. No other instrument has revealed as much to man about the world he lives in. And with the gathering of rocks and dust from the moon the microscope is helping man to find out more about worlds other than his own.

The term "light microscope" does not refer to weight but to the fact that the instrument uses light to provide a magnified image of the object under view. Light microscopes were used over two hundred years before the first electron microscope was made.

It would seem that the inventor of so important an instrument as the light microscope would have an honored place in the history of science. No one person, however, can be named as the inventor of the microscope. Its development can be traced to the first observation that looking through clear substances of certain shapes would result in magnified images.

Anton van Leeuwenhoek is often credited with being the "inventor" of the microscope. Leeuwenhoek did indeed make a number of single-lensed microscopes around the 1670s, but this Dutch cloth merchant did not make the first microscope. It is known that, as early as 1590, two brothers, Hans and Zacharias Janssen, also in Holland, made a microscope by placing lenses at the opposite ends of a tube. Since the Janssen instrument had more than one lens, it was a compound microscope. As such, it was more like modern microscopes than Leeuwenhoek's was. There were probably instruments that could be described as microscopes before 1590, but there is no clear record of them.

Leeuwenhoek's microscopes were single lenses mounted on a flat metal plate. The material to be viewed, usually a drop of liquid, was suspended on a needle-like affair on one side of the plate. The entire instrument was brought up to the eye and the specimen viewed through the lens. The needle could be moved to get the best focus. A single-lensed microscope such as Leeuwenhoek's is termed a simple microscope.

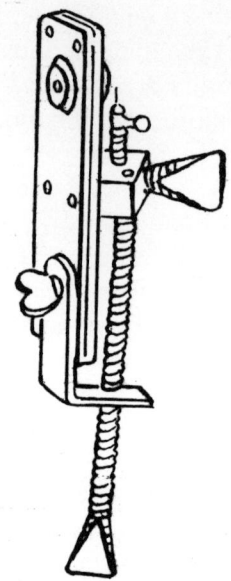

An artist's rendition of a Leewenhoek microscope; the specimen was placed on the needle-shaped rod.

Some ten years before Leeuwenhoek made his microscopes, an Englishman, Robert Hooke, used a microscope of compound design. Hooke's microscope was a beautiful piece of workmanship that could achieve magnifications of about thirty diameters. Hooke made many observations of commonplace objects with his microscope, and the commonplace became objects of incredible wonder and beauty. As seen through Hooke's microscope, the stinger of a bee became a viciously barbed harpoon. A fly's wing became a gossamer landscape crisscrossed with branching rivers.

Hooke's work did much to stimulate interest in microscopy. The microscope revealed a world the existence of which had never been suspected. Some of the most remarkable observations, however, were made by Leeuwenhoek with his single-lensed instruments. He saw tiny living things which he called "little animalcules" and "cavorting beasties." Leeuwenhoek's animalcules and cavorting beasties are today called bacteria, protozoa, and fungi, among other things. Leeuwenhoek was able to achieve higher magnifications with this single lens than Hooke could with his compound microscope. Judging from his descriptions and drawings of bacteria and protozoa, his lenses probably magnified about 300 to 400 diameters. Leeuwenhoek was very secretive about his instruments and would tell no one how he managed to make his tiny lenses. Fortunately, he was not as secretive about his observations, and he left many detailed records.

Hooke put together a book based on the work with his compound microscope. This book, entitled *Micrographia*, was illustrated with many drawings that Hooke made directly from his observations. Perhaps the best known of Hooke's observations are those he made of thin slices of cork. In his description he noted that cork "consisted of great many little boxes. . . ." And he referred to these boxes as "cells." That all living things are made of cells is a basic axiom of biology. Hooke certainly did not formulate the cell theory, and he had no idea of the real significance of his observations. But the name he applied to the cork "boxes," which were the empty remains of dead cells, is now used for the living thing.

The compound microscope, such as that used by Hooke, was destined to evolve into the modern microscope. Almost all microscopes in use today are compound microscopes. There are many different kinds of light microscopes designed for many different purposes. Most microscopes are designed to view specimens with transmitted light. In these instruments the light passes directly through the specimen, rather than being reflected from it. Specimens viewed with these microscopes must necessarily be quite thin. There are some microscopes—especially those used by metallurgists, the scientists who study metals—that are designed to view specimens with light reflected from the surface of the specimen.

Basic to a compound microscope are a source of light; an objective lens; an eyepiece lens, or ocular; a place to put the specimen so that light can get at it or through it; and a means of changing the distance between the objective and the specimen so that the sharpest focus or clarity of image can be obtained. Actually the objective "lens" and "eyepiece lens" are both made of many lenses. There are, of course, many other things that can be put on a microscope to increase its usefulness and add to the convenience of the user. Most compound microscopes also have a condenser lens that is placed between the light source and the specimen. The condenser lens concentrates the light that passes through the specimen. A microscope usually has a platform, called the *stage*, on which the specimen is placed. Specimens are usually mounted on glass slides. Another thin piece of glass, called a *cover slip*, is frequently placed over the specimen. The light passes from the source to the specimen through a hole on the stage. The microscope usually has a device for making this hole larger or smaller, thereby increasing or decreasing the amount of light that passes through the specimen. The amount of light can also be varied by a dimmer switch that increases or decreases the light at the source. There are frequently clips on the stage for holding the glass slide. More expensive microscopes have what is called a *mechanical stage*, a device that moves the slide by means of turning knobs.

Magnification is changed by changing the objectives and ocular. Most microscopes are equipped with several objectives on a revolving turret. The objective can be changed by swinging another into place. Other microscopes have *zoom* oculars that can achieve different magnifications by a knob that changes the lens arrangement.

When the microscope is used, light passes from the source through the condenser and specimen into the lenses of the objective. The light passes from the objective into a tube and into the ocular mounted at the opposite end of the tube from the objective. The purpose of the tube is to hold the objective and the ocular at the right distance from each other and to keep out unwanted light. From the ocular the light goes to the eye of the observer. The objective lens forms a magnified image and the ocular magnifies the image further. The total magnification is the product of the magnifications of the objective and the ocular. For example, if the objective magnifies 40 times and the ocular magnifies 10 times, the total magnification is 400.

Of course, there will be no image unless there is a specimen. As the light passes through the specimen, the light is affected according to varying thicknesses and colors in different parts of the specimen. This interference in the passage of the light results in the pattern that is the image or picture of the specimen.

High magnifications are obtained by combining a variety of lenses of different shapes and types of glass in the objectives and oculars. As magnification is increased, however, the area of the specimen which can be viewed at one time goes down.

Early microscopists had to contend with distorted images and unwanted rings of color in the images. The distortion was a matter of lens quality and was eventually solved by improving lenses and discovering better ways of combining lenses. The color was a *prism effect* and was due to the physical fact that when white light passes through a medium such as glass at certain angles, it tends to be broken down into the colors of the spectrum or rainbow. This bothersome effect is called *chromatic aberration.*

A breakthrough in achieving clear high magnification came in 1877 with the invention of the "oil-immersion" objective by Ernst Abbe, a German microscope maker. Abbe was also the inventor of the condensor lens. The oil-immersion objective is composed of a lens system designed to achieve the maximum refraction for high magnification. The tip of the objective is immersed in a drop of clear oil that has been placed on the cover slip. The result is a clearer image as the additional refraction in the air space between the objective and specimen is eliminated.

With the use of the oil-immersion objective, meaningful high-magnification observations were possible, and by the turn of the century many microscopists were somewhat surprised to find that the upper limit of magnification with light microscopes had been reached. No combination of lenses, oil immersion, and application of techniques could achieve magnifications much higher than about 2,000 diameters. There was much to be seen that required higher magnifications—viruses, for example. That the limit of magnification had been reached did not mean that the development of the light microscope had come to an end. Many more improvements in microscope design were to be made, and the light microscope is still being improved. The phase-contrast microscope, for example, enables biologists to investigate the internal structure of living cells in far more detail than was possible with microscopes of 1900. The phase-contrast microscope has parts that alter the characteristics of the light before and after it passes through the specimen. The result is that the internal structure of nearly transparent specimens can be seen in great detail without staining. But the microscopes of 1900 achieved magnifications just as high as present-day microscopes do. It was clear that if the 2000-diameter barrier was to be broken, it would have to be with something other than light.

The development of electron microscopes followed from the use and improvement of light microscopes. Research into the nature of light resulted in a chain of discovery that eventually uncovered the knowledge that made electron microscopes possible. The use of light microscopes pointed up the need for an instrument that could magnify more and enable scientists to see more than they could with light microscopes.

# 2
# A New World Seen and Unseen

The use of improved microscopes laid the very foundation of biology, and the use of the microscope was extended to other sciences. With the aid of the microscope the cell theory was firmly established. It was discovered that all living things began life as one cell and that life could come only from preexisting life. The details of the fertilization of egg cells by sperm cells were observed as well as the details of embryonic development. The minutiae and detail of the structure of living things, from the largest to the smallest, were noted. Man was made healthier, and his span of life increased because of the use to which scientists put this remarkable instrument. It is well known today that many of the tiny organisms Leeuwenhoek called animalcules and beasties can cause disease in man and other living things.

The connection between microscopically sized organisms and disease was not immediately apparent to Leeuwenhoek and other microscopists who came after him. Some 200 years were to pass before firm, scientific evidence of the bacterial causes of disease was offered. During those 200 years there were many who suspected that certain bacteria would certainly do no good if they got into an open wound or were inhaled or eaten. But there were as many more who dismissed as ridiculous the idea that bacteria could cause disease. They argued that the microscope had made it obvious that these little things were everywhere in great abundance and that if they caused disease, everyone would die after taking his first breath. Of course, at the time, little if anything was known about immunity and natural resistance to disease organisms. With the use of the microscope, rod-shaped, spherical, and spiral-shaped bacteria were in fact observed in the blood and tissues of sick animals and people. But it was argued that they were also observed in healthy animals.

The foundation of *the germ theory of disease* was established in 1865 by a French chemist, Louis Pasteur, and it was sour wine, not sick people, that gave him the clue. Pasteur was called upon to investigate the souring of wine at some French wineries. Examining normal wine with the microscope, he found roundish yeast cells. But this was not a particularly spectacular observation. Everybody knew that yeast cells were always found in wine. Upon microscopic observation of sour wine, he found, not yeast, but rod-shaped bacteria. Further experimentation revealed that it was indeed the bacteria that caused the wine to sour and, furthermore, that it was the yeast that caused grape juice to turn into wine. Pasteur reasoned that if bacteria could cause wine to sour, then these little *microbes* (as Pasteur called them) might wreak havoc in the form of disease if they got into man or some other animal or plant.

Pasteur's work started a new science, bacteriology, and the microscope was the key instrument in this new science. Thousands of scientists followed Pasteur's lead and turned to bacteriological studies. So much work was done in bacteriology from the 1880s to the early 1920s that this period has been called the Golden Age of Bacteriology. Among the more outstanding of the bacteriologists of this period was Robert Koch. Koch discovered, among other things, the bacterial causes of tuberculosis and anthrax, a disease of sheep and sometimes of man.

It was through the anthrax work that Koch devised the basic method of proving that a particular microbe was the cause of a particular disease. Using the microscope, he found rod-shaped bacteria in the blood of a mouse that had died of anthrax. He isolated the bacteria by growing it on an outside growth-medium. Koch used the sterile fluid from the eyes of oxen as his growth medium. He collected the fluid into containers and introduced the bacteria into it. The organisms thrived in the fluid. Today the usual medium for growing bacteria is a gelatinous material called *agar.* Koch then infected some of the bacteria he had grown into healthy

An early electron microscope manufactured in 1939 by the Siemens Aktiengesellschaft Company.

mice, and the mice, developing symptoms of anthrax, died. Microscopic examination of the blood of the dead animals revealed huge numbers of rod-shaped bacteria, the same kind of bacteria that had been injected into the mice. The bacteria had multiplied in the blood of the mice and were quite obviously the cause of anthrax. The evidence was made more conclusive with further experimentation, using large numbers of mice.

Other bacteriologists used Koch's methods to find causes of many more diseases. And when the causes were discovered, specific drugs to effect cures and vaccines for prevention could be developed. Many bacteriologists, however, were quite dismayed on those occasions when Koch's methods would not work. Microscopic examination of the blood of animals who were very sick or dead of a disease sometimes revealed no microbes, or at least no microbes that could be demonstrated as the cause of the disease under investigation. Among these ailments were rabies, yellow fever, and measles. Even use of the best quality microscopes at the highest possible magnification (about 2000 diameters) did not enable the bacteriologists to see what caused rabies, yellow fever, measles, and many other diseases. But there was something that was causing the disease; for if blood, or some other body fluid from an animal sick with one of these afflictions, was introduced into a healthy animal, the healthy animal developed symptoms of the disease.

Bacteriologists had at their disposal a porcelain filter that was used for separating bacteria from fluids. It was something like a superfine colander with holes so tiny that even the smallest of bacteria could not get through. The mysterious disease-causing material that could not be seen went right through these filters. And for this reason it was, for many years, called a *filterable virus*. The word "virus" means "poison," and for many years it was used to designate anything bad, foul, or noxious. Many thought that since the filterable virus went through the filter, it had to be a liquid; it was also thought that the liquid itself caused the disease rather than some "solid" particle, such as bacteria. Other bacteriologists thought that the virus was a particle so fantastically small as to make most bacteria gigantic by comparison. Viruses further confounded bacteriologists by refusing to grow in any of the usual growth media.

As the twentieth century began, no one had knowingly seen a virus. In 1931 William Elford devised a filter with pores small enough to trap the elusive viruses. This development pointed out that viruses were particles rather than liquid. But this only increased the frustration of bacteriologists who then knew that there was something to see, even though they still could not see it with the microscopes they had.

Viruses were isolated in 1935 by Wendell Stanley in the course of working with a disease of tobacco plants called tobacco-mosaic disease. The name of the disease came from the mottled ap-

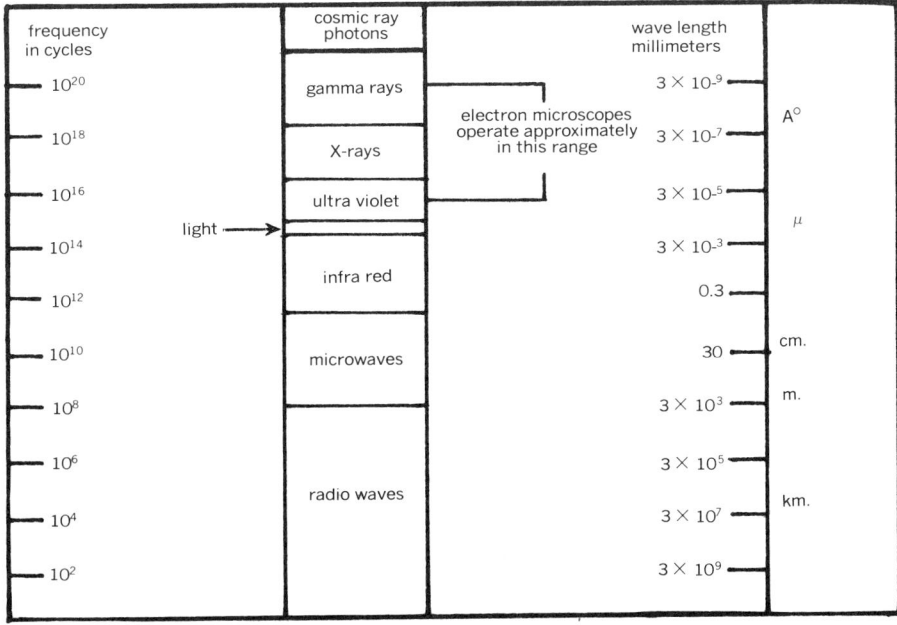

pearance of infected plants. From about a ton of infected tobacco leaves Stanley extracted less than a tablespoonful of a white, crystalline powder. The powder appeared to be no more alive than a pinch of table salt. But the powder was indeed the virus, for if it was rubbed into a tobacco leaf, the leaf soon showed symptoms of the mosaic disease. Viruses seemed to be alive only when they got into another living thing. Microbiologists were now more anxious than ever to see viruses.

The frustration of not being able to see something was not limited to scientists who studied viruses. Toward the beginning of the twentieth century it was obvious that all branches of science were in need of more powerful microscopes. Others, such as biologists who studied the structure of cells, could see tiny inclusions in the cells they observed, but even the best of microscopes revealed them as only tiny, hazy specks. Physical scientists, geologists, and technologists who studied rocks, metals, fabrics, and other materials were also in need of higher-powered microscopes. The highest-powered light microscopes could not reveal to a metallurgist the details of structure that made one alloy stronger than another. And lacking this knowledge, the development of new alloys was largely a guessing game. Electron microscopes did not become available until the 1930s, but the process of acquiring the knowledge that would make them possible had started as early as Robert Hooke's day among scientists who studied the nature of light.

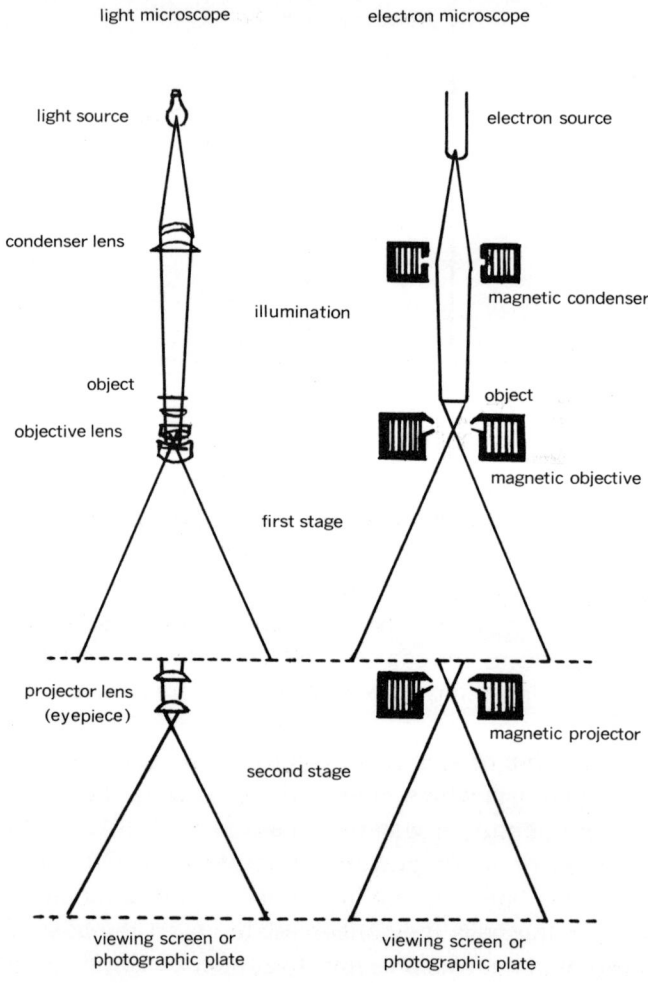

A comparison of the pathways of light and electron beam in the light and electron microscope.

# 3
# From Light Microscope to Electron Microscope

Light is, of course, what enables one to see. This one fact alone has made light fascinating and an object of study of many scientists. One of the first scientists to study light seriously was Sir Isaac Newton, who lived about the same time as Robert Hooke. Even before Newton it was known that light traveled in straight lines. Everybody knows that an opaque object in a light path will cast a shadow. This is because the light, traveling in straight lines, will not curve around the object. Among other things Newton determined that white light is made up of all the colors of the spectrum. Newton discovered this by passing a beam of light through a triangular piece of glass called a *prism.* The light that emerged from the prism cast a band of color on a piece of paper placed in the light path. Newton's prism experiments also tended to verify a characteristic of light that was already being put to use in microscopes. The light emerged from the prism at an angle to the path the light followed on entering the prism. This would appear to be a contradiction to the statement just made about light traveling in straight lines. Light does indeed travel in straight lines as long as it travels through the same medium, the medium being the substance through which the light passes. Most of the light that enters our eyes has traveled through the medium of air. When light passes from one medium to another at an angle, the speed of the light is changed, and the light is bent. If the light passes from a thin medium, such as air, into a thicker one, such as glass, the light

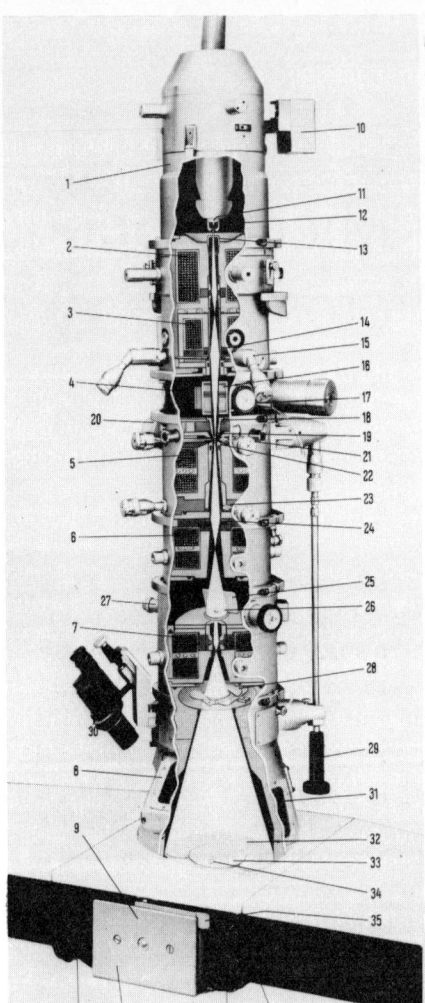

1 Electron gun
2 Condenser 1
3 Condenser 2
4 Specimen airlock with beam deflecting system
5 Objective
6 Intermediate lens
7 Projector
8 Final-image tube
9 Photographic chamber
10 Servomotor for cathode displacement
11 Cathode
12 Bias shield
13 Anode
14 Electromagnetic stigmator in condenser 2
15 Control for the apertures in condenser 2
16 Deflecting system
17 Automatic ventilation and evacuation of the airlock antechamber
18 Specimen cartridge
19 Specimen
20 Anticontamination device
21 Control for the objective apertures
22 Electro-magnetic objective stigmator
23 Intermediate lens selector apertures
24 Electro-magnetic intermediate lens stigmator
25 Mirror for intermediate image observation
26 Intermediate image screen
27 Drive for the intermediate image screen
28 Exposure shutter
29 Control for specimen stage adjustment
30 Binocular magnifier
31 Viewing window
32 Outer field screen
33 Central field screen
34 Small field screen
35 Lever for opening the photographic chamber door
36 Control for final-image screen (33 and 34)
37 Drive for film and plate transport
38 Door of the photographic chamber
39 Airlock drive

Fig. 1/2 Microscope column, longitudinal cross section

A cutaway view of a modern transmission electron microscope with the major parts identified.

is slowed down. When light passes into a lens, it is passing into a thicker medium, and the light is bent. Another word for "bend" when applied to light is "refract." It is the nature of the refraction that determines whether looking at something through the lenses will give you more information about that particular thing than looking at it with the eye alone. The lenses of a microscope are made so that the light is refracted in a way that results in a clear, highly magnified image.

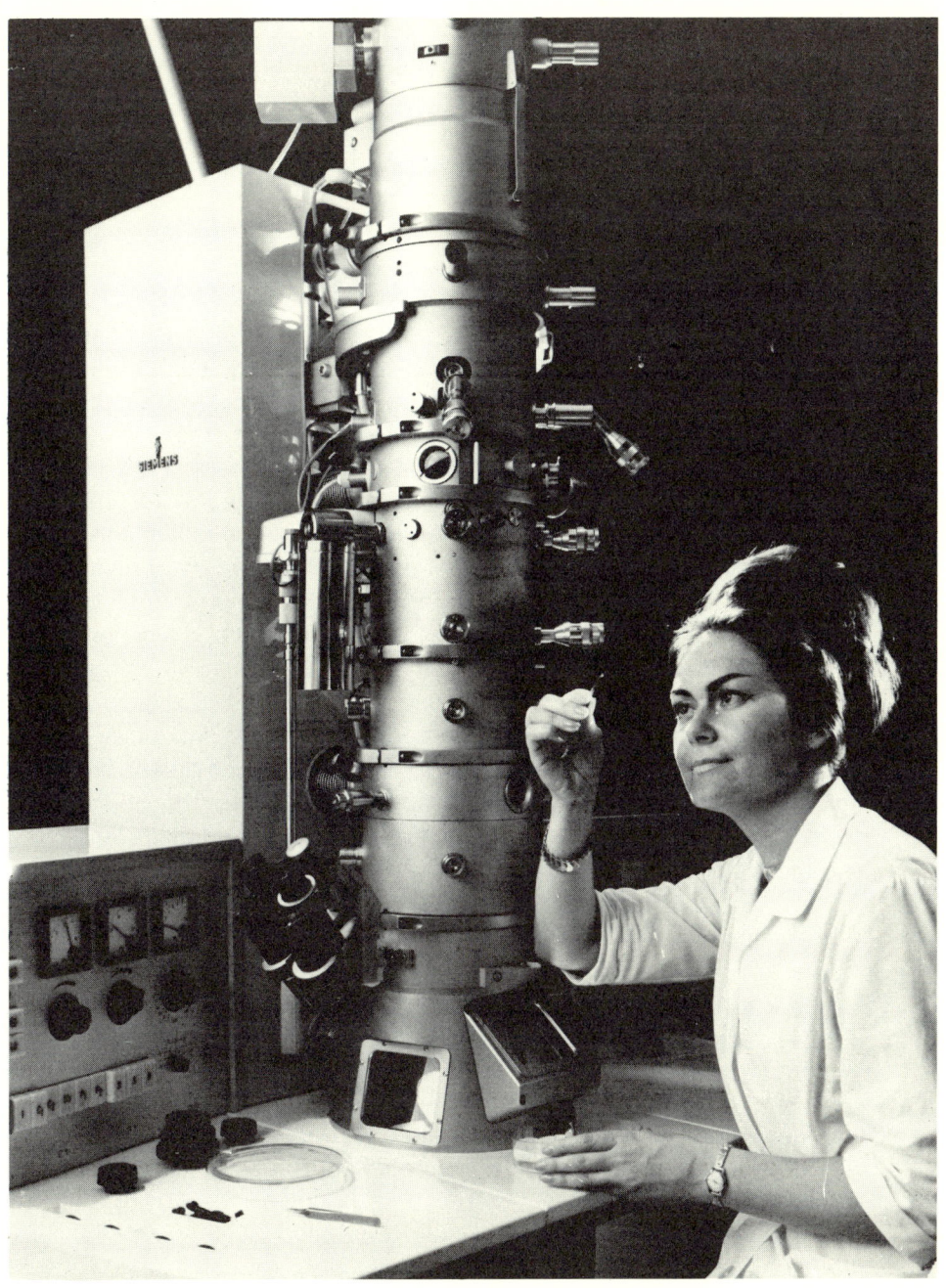

A modern transmission electron microscope.

Refraction can be demonstrated by putting a pencil or some other similarly shaped object into a glass of water. If the pencil is viewed from the side of the glass of water, the pencil will appear to be bent or broken. This is because the light entering the glass of water from the air is passing into a denser medium and is refracted.

The refractive behavior of light can be put to good use with lenses. A lens is usually made of glass, but it can be made of any transparent material such as plastic or even water. Indeed, the first lenses were probably water. Dew drops can act as lenses. The Roman philosopher Seneca wrote in A.D. 65 that glass goblets filled with water were useful for seeing objects "that frequently escape the eye."

Lenses are made in a great variety of shapes and sizes and what happens to light when passing from air into the medium of a lens depends upon many factors, such as the shape of the lens and the material of which the lens is made. One thing that always happens, however, is that the light is bent or refracted. The result to the eye of the observer can be magnification, reduction, or just distortion. And it is the skill of lens grinders that makes looking at things through lenses a meaningful experience.

The most familiar kind of lens is the common magnifying glass. If the magnifying glass is held between the viewer and an object, the object appears to be bigger. Just how much bigger the object appears to be depends on the shape of the magnifying glass lens and the distance the refracted light travels from the object through the lens and to the eye of the observer.

Resolving power is an important quality of a microscope. The resolving power of a lens or lens system is the smallest distance between two points that can be clearly separated to the eye of the observer. When the resolving power is exceeded, the image appears blurred, the magnification is "empty." The resolving power is actually just as important if not more so than magnification in a microscope or any other lens system. For magnification is useless if what is magnified cannot be clearly seen.

Leeuwenhoek's simple microscopes were actually the same thing as a magnifying glass. Leeuwenhoek patiently ground every one of his tiny lenses to the right shape and curvature to achieve high magnifications and, apparently, high resolution. He could not

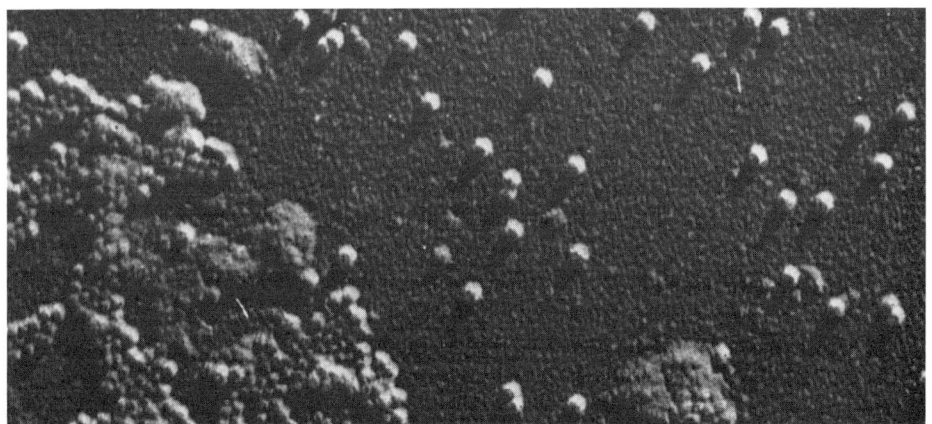

First verified electronmicrograph of poliomyelitis (polio) virus 65,000X; the "shadows" are areas not coated by the vacuum evaporation process (see page 64).

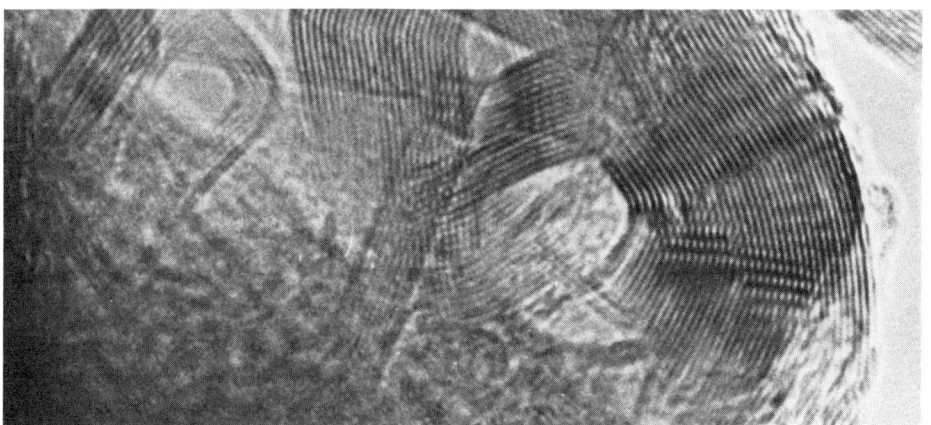

Transmission electronmicrograph of graphitized carbon black (280,000X).

have described his observations as well as he did unless he had fairly good resolution.

The knowledge that light travels in straight lines and that it can be bent or refracted explains the basic way in which lenses work. This knowledge, however, still does not explain why light microscopes are limited to 2000-diameter magnification. This was not explained until more was known about the wave motion of light and other radiation.

Newton again started this train of discovery. He thought that light was made up of little particles he called *corpuscles* Other scientists did not agree with this, maintaining that light consisted

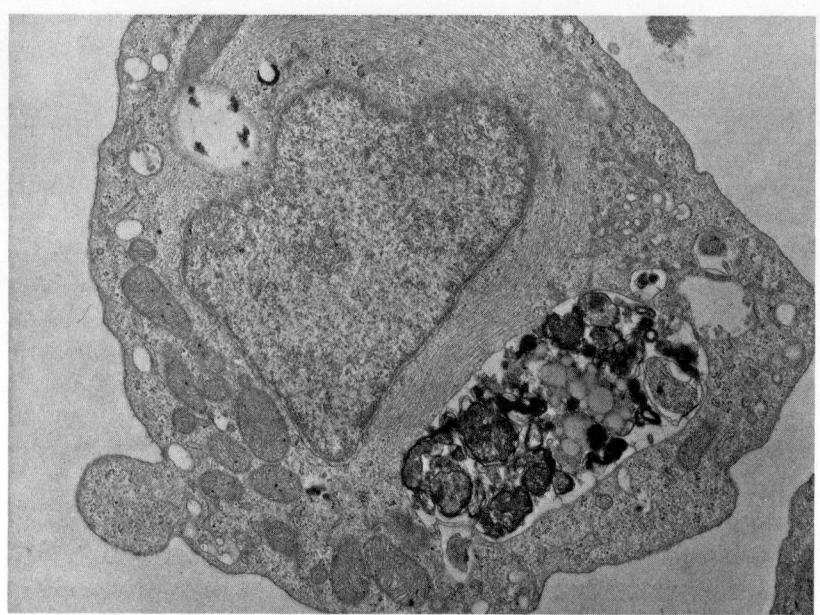

A white blood cell-transmission electronmicrograph at 25,000X. The darker structure at lower right is an object, probably a bacterium, that the cell has ingested. The somewhat heart-shaped structure is the nucleus. The double nuclear membrane can be seen. Endoplasmic reticulum is around the nucleus.

of fine waves. Notable among these was the Dutch physicist Christian Huygens, who proposed his wave theory in 1678. Huygens also designed a type of eyepiece that is still widely used in microscopes and telescopes. Physicists argued over the corpuscle and wave theories for more than a hundred years. The wave theory was strenghthened considerably by the work of the English physicist Thomas Young in 1801. Young constructed a small, upright partition with two closely spaced holes. He directed a narrow beam of light through the holes at a screen placed behind the partition. If light consisted of particles, some of the particles would merge as they passed through the holes and would produce a brighter image on the screen where they overlapped and a dimmer image where they did not. What he observed on the screen was a series of bands of light separated by intervals of darkness.

Young's observation could be explained by the wave theory. The bright areas represented waves reinforcing each other somewhat like the crests of waves hitting on a beach together. Such crests of waves are said to be "in phase." When waves are out of phase, the trough of one wave cancels out the peak of another. Young proposed that the dark bands represented waves striking the screen when they were out of phase. Therefore, since they canceled each other out, there was little or no light energy to strike the screen at those areas.

The wavelength of light was determined by Young's calculations based on the ratio of the width of the bands to the distance between the holes. The wavelengths that constitute visible light turned out to be very small. They were so small that eventually new units of measurement had to be defined to deal with them conveniently. These small units are generally used today to express the size of things that are so small that they can be seen only with an electron microscope.

The millimeter is, in terms of everyday types of comparison, a small unit of length. It is $1/1000$ of a meter and, expressed in terms of the English system, a millimeter is about $1/25$ of an inch. The millimeter, however, is much too clumsy for use in most microscopic work. For expressing the size of things such as bacteria or protozoa, the micron was defined. A micron is $1/1000$ of a millimeter. It is represented by the Greek letter *mu* ($\mu$). Even the micron, however, is too cumbersome for such purposes as measuring the wavelengths of light and the size of objects generally observed with electron microscopes. For measurements in this realm the unit used is the Angstrom, named after the Swedish astronomer, Anders Jonas Angström, who proposed it. The Angstrom, represented by a capital A, is $1/10,000$ of a micron, or 0.0000001 mm., or 0.00000025 inch.

The wavelength of visible light ranges from about 3500 A to about 7500 A. Our perception of color is dependent upon the wavelength of the light striking our eyes. The shorter waves at the 3500 A range are perceived as violet. The longer waves at the 7500 A end are seen as red. The other colors lie between 3500 A and 7500 A. Wavelengths from about 5600 A to 5900 A are perceived as yellow. Green is represented by a fairly wide range from about 4900 A to 5600 A.

The determination of the extreme shortness of the wavelengths of light cleared up many things about its behavior. The small wavelength prevented light from going around any object that was larger than the wavelength of the light. It also explained to some extent how microscopes worked. Although bacteria and other

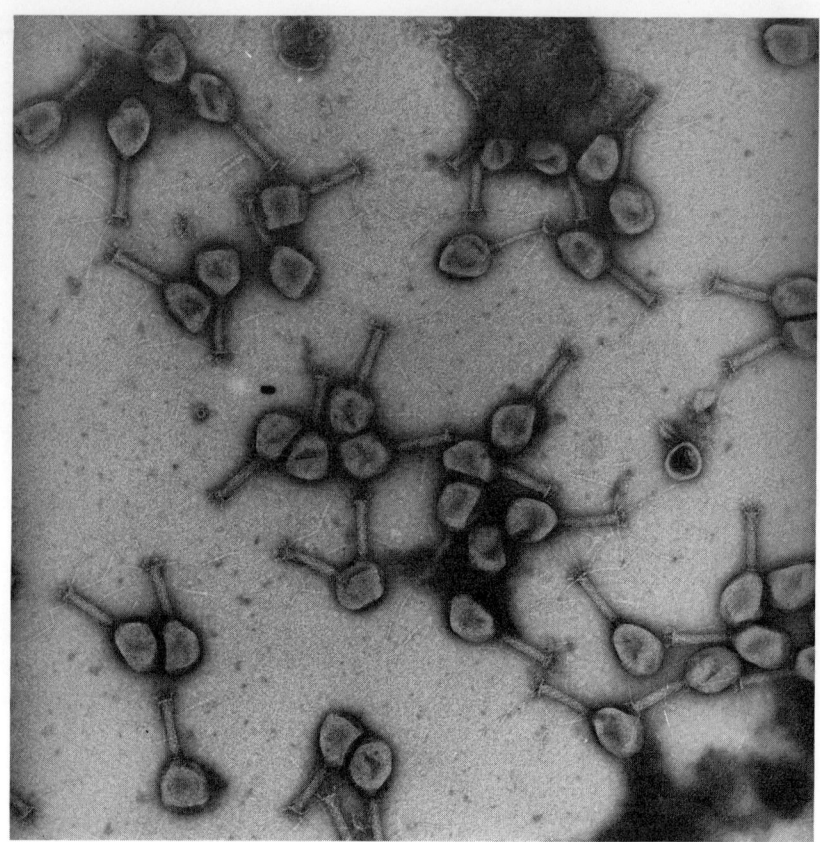
T-4 bacteriophage viruses. Transmission electron micrograph at 200,000X.

things viewed with microscopes are exceedingly small compared with anything that can be seen without a microscope, they are enormous when compared to wavelengths of light. A typical bacterium is about 500,000 A in length. Light waves cannot go around them, so light can define their presence.

If an object is smaller than light wavelengths, the light will go around it, so no amount of refraction with lenses will make the objects visible. An object with a size near that of light wavelengths, (or a distance between objects) cannot be defined sharply enough to give clear resolution. As the magnification is increased, resolution is lost. And the point where resolution is completely lost with the best of light microscopes is about 2000X.

A great surge of scientific work started around the last quarter of the nineteenth century. Thousands of biologists peering into improved microscopes observed and described the minute details of cell structure, cell dimension, fertilization, and embryology. Much of the pioneer work, discussed earlier, in discovery of disease organisms was carried out during this period of time.

Developments in physics at this time were also very significant. Among other things radio waves, X rays, and various other such radiations were discovered. It was found that these radiations, as light, traveled in waves at the same speed as light. Furthermore, it was found that these radiations were all essentially the same kind of energy. They were called electromagnetic radiation. Eventually a wide range of electromagnetic radiation was identified. The major difference between one type, such as light, and another, such as radio waves, is the wavelength. Most electromagnetic radiation comes from the sun. But a variety of man-made devices, such as radio transmitters, X-ray machines, and light bulbs produce electromagnetic radiation. Physicists found that some of these waves could be bent (refracted) by magnetic force and by certain minerals such as quartz.

By the last quarter of the nineteenth century the theory of the atomic structure of matter was well accepted. It was known that atoms combine in definite proportions to form molecules of various substances. No one was quite sure just why the atoms came together or what held them together. Nor did anyone at the time know anything about the structure of the atom.

The study of electricity, apparently quite unrelated to chemistry, in the 1860s, led to the formulation of a picture of the atom. The curiosity of physicists as to what would happen if electricity were sent through a vacuum led to information on atomic structure and ultimately to electron microscopes and other electronic instruments. This was made possible by the invention of a glass tube that could be pumped out and then sealed with the electrodes in place. When the current was sent through the tube, a greenish glow was noted on the wall opposite the negative electrode or cathode. Some sort of radiation was coming out of the cathode, and when this radiation struck the wall of the tube, the green glow resulted. The radiation was called a *cathode ray.* Later, when the English physicist, William Crookes, improved the tube it was called a *Crookes tube.* The big question that, no doubt, provided physicists of the time with hours of stimulating debate was: Were cathode rays waves or particles?

In 1897 the British physicist Joseph John Thomson demonstrated that cathode rays could be deflected from their path by magnets or electric charges. This indicated that they were particles, and the nature of the deflection showed that they were negatively charged particles. The particles were called *electrons*, and it was soon evident that the electron was a subparticle of the atom. Up to that time, atoms had been thought to be indivisible which is why they were called atoms (from a Greek word that means "indivisible"). In the next century more atomic subparticles were discovered and a picture of the atom evolved. The picture or model of the atom is subject to many interpretations. The basic idea, however, is of negatively charged electrons spinning in orbits or "clouds" around a central nucleus. The major nuclear particles are positively charged *protons* (equal in number to the electrons) and neutral particles called *neutrons.* Many more types of nuclear particles have been found and are still being found. The electron was also identified as the basic unit of an electric current. The mass of the electron was determined to be about $1/1837$ of the mass of a hydrogen atom.

At the start of the twentieth century the wave nature of light and other electromagnetic radiations and the particle nature of electrons were generally accepted. But the work of a German physicist Max Planck completely changed the direction of physics. It came about through some questions raised by the emission of radiation from heated bodies of matter. Much of, but not all of, this radiation is light. Mathematical calculations involving radiations at both ends of the visible spectrum showed some inconsistencies that were not explained by wave motion.

In the late 1890s Planck proposed some calculations (equations) that indicated that radiation consisted of tiny packages of energy which he called *quanta*, from the Latin word for "how much." Planck proposed that radiation could be absorbed or given off only in whole numbers of quanta. The amount of energy in a quantum depended upon the wavelength of the radiation; the shorter the wavelength the more energy in the quantum.

The quantum theory was used by Albert Einstein and others to explain other phenomena. Einstein proposed the term *photon* for a quantum of light. Einstein used it to explain the *photoelectric effect*, which is the emission of electrons when certain metals are struck by light. The photoelectric effect is the basis of the photographer's light meter. Einstein proposed that in some situations electrons absorbed enough energy from quanta of light to leave the surface of the metal.

The quantum theory is still in good repute with physicists, although to many nonscientists it does not make sense. It seems to be similar to Newton's corpuscular nature of light. Waves can have the properties of particles and particles can be like waves. It is sometimes convenient, for example, for physicists to think of light as waves and at other times as particles.

In 1923 Louis de Broglie, a French physicist, offered some mathematics that implied a dual wave-particle nature for electrons in orbit and in so doing established the theoretical basis for the electron microscope. De Broglie suggested that electrons in motion were guided by waves that were associated with the motion. He even predicted the wavelengths of these electron waves, calculating that they would be about 2 Angstroms. De Broglie's predictions were upheld by experiments carried out in 1927 by Clinton Davisson and Lester Germer of the Bell Telephone Laboratories. They passed electrons through crystals of the metal nickel and then directed the electrons onto photographic film. Interference patterns were observed on the film and this verified the wave nature of electron beams. The wavelengths were determined by measuring the interference bands and were found to be 1.65 A, very close to de Broglie's prediction.

The 1.65 A wavelength was fantastically smaller than the 3500 A to 7500 A range of visible light. Waves can be refracted, a discovery that became the basis of the electron microscope. It was then obvious that if a way could be found to direct electron beams through a specimen and focus them onto a photographic plate or onto something that would convert the electron beams to visible light, a microscope of incredible magnifying and resolving power could be made. Some ten years later the first electron microscope was a reality.

Even before Davisson and Germer proved the wave nature of electrons the possibility of making an electron microscope had not gone unnoticed. De Broglie's electron-wave hypothesis had encouraged physicists to look for ways to focus electron beams. One year before the work of Davisson and Germer, H. Busch, a German

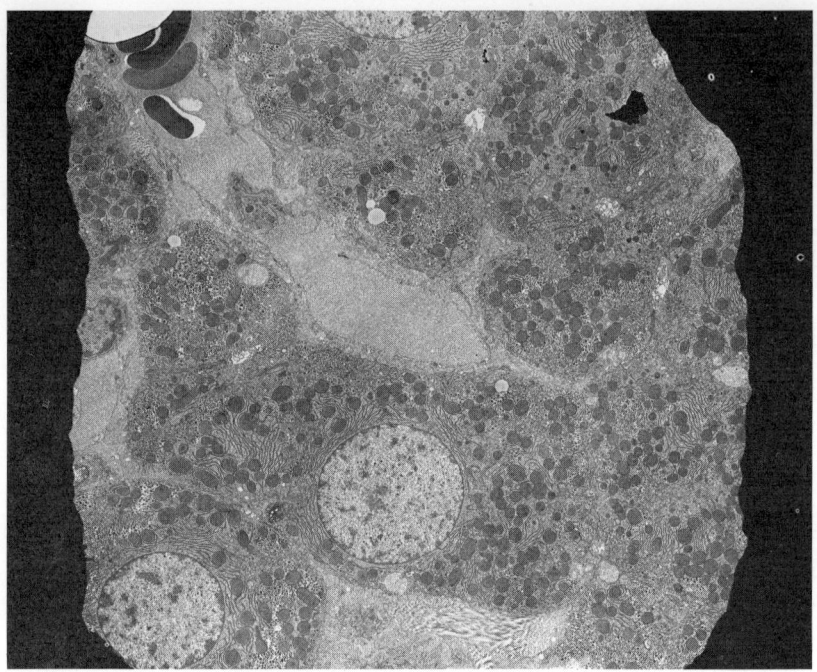
Transmission electronmicrograph of rat liver cells (4,500X).

physicist, proposed that magnetic or electric fields could act as lenses for electrons and other charged particles. Busch had been working on the effects of magnetic fields on charged particles for some fifteen years before De Broglie presented his ideas. By 1927 Busch and other physicists, notably another German, Ernst Ruska, had actually made magnetic electron lenses which would refract beams of electrons.

The idea of a magnet acting as a lens may seem strange when thought of in relation to a glass lens. It is not, however, the magnet itself that is the lens. Rather it is the magnetic field produced by the magnet that acts as a lens for electrons. The operation of a magnetic lens is based on the principle discovered in 1897 by J. J. Thomson that electron beams in a Crookes tube could be bent by magnetic fields.

Ruska, in collaboration with R. Knoll, continued to work on improving magnetic electron lenses. They tried permanent magnets and electromagnets and soon found that best results could be obtained with electromagnets. Electromagnets are basically insulated wire wrapped around an iron core. Almost every school child has made a simple electromagnet by coiling some insulated wire around a nail. When the ends of the wire (stripped of insulation) are connected to a battery, the nail becomes a magnet. The electromagnets made for use as lenses in electron microscopes are, of course, more complicated than some wire wrapped around a nail, but the same basic principle is at work.

Electromagnets are better for electron microscope lenses than permanent magnets mainly because the electromagnets can be controlled. More electricity results in a more powerful magnetic field and the field can be changed to achieve the desired focusing of the electron beams. In that respect magnetic lenses are more flexible than glass lenses. The magnifying power of magnetic lenses can be changed by altering the amount of electric current that goes into the wire windings. To change the magnification of an optical microscope, the operator has to change lenses, or in the case of zoom microscopes, move the lens elements.

Knoll and Ruska's basic design was a winding surrounded by an iron housing. There was a gap in the housing. The gap allowed the electron beams to enter and be refracted by the magnetic field.

Ruska in 1934 reported that he had developed the basic design of an electron microscope. In 1937 Knoll and Ruska demonstrated a device that was essentially an electron microscope. In 1931, however, a man named Rudenberg, an employee of the German Siemens Company, had applied for a patent on what he called an electron microscope. So, although Knoll and Ruska are frequently credited with being the inventors of the electron microscope, their claims are by no means undisputed. Ruska and another German physicist designed an improved electron microscope for the Siemens Company in 1939. It could achieve a resolution of 100 A.

Early electron microscope work was not limited to Germany. In Canada a practical working electron microscope was made in 1939 by A. Prebas and James Hillier. The first electron microscope offered for sale was made in the United States by the Radio Corporation of America (RCA) in 1940. James Hillier contributed to its development. It has a resolving power of 24 A. Ruska in the meantime had improved the instruments he was making for Siemens to a resolution potential of 10 A. These instruments, and most of those in use today, are transmission electron microscopes. They are called this because the electron beam is transmitted, that is, passes through the specimen, as light passes through the specimen in most light instruments.

Further development of electron microscopes was delayed by World War II. Very soon after the war, however, development started again, and electron microscopes were soon a multimillion-dollar business. Many companies entered the field, and the pressure of competition stimulated rapid improvements in its design and capabilities. By the 1960s electron microscopes were capable of resolutions of 1 A or better in good conditions, and magnifications of 200,000 diameters and more. This betterment of the 1.65 A wavelength, as determined by de Broglie, was made possible by high voltages which reduce the wavelength of the electrons.

Now the range of observation available to scientists was dramatically extended. Parts of cells that had been only hazy specks were clearly seen as fantastically elaborate structures. Viruses came to be a favorite target of electron microscopists, and the electron microscope revealed virus particles to have remarkably precise geometric shapes.

Electron microscopes became generally available before much was known about how to use them or how to prepare the specimens for viewing. Both instrument operation and specimen preparation require much more training than is the case with light microscopes. Electron microscopes have some limitations. With very few exceptions living material cannot be examined. Beams of electrons are scattered by air so the path of the beam and the specimen must be in a vacuum. The image formed by the electron microscope is black and white. The electron beam does not have the wavelengths that we perceive as color or perceive at all for that matter. The image is formed in somewhat the same way as the image of a black and white television set by the electron falling on a fluorescent screen. Even with these limitations, the electron microscope has extended the observable world down to the very molecules, and in some instances, the atoms that compose the living and nonliving world.

# 4

# Transmission Electron Microscopes

For all the difference in appearance and methods of operation the optical microscope and the electron microscope are quite similar in many ways. Recall that in the transmitted-light microscope light passes through a condenser lens which intensifies it. The light then passes through the specimen into the objective lenses where it is refracted, then to the eyepiece or ocular, which further magnifies the image, and then to the eye of the observer.

The path of the electrons in an electron microscope is similar to the light path in an optical microscope. Electrons stream from a source into a magnetic condenser where they are intensified. The electrons then pass through the specimen into magnetic objective lenses, where it is refracted, then into a magnetic projector lens which projects the electron microscope on to a fluorescent screen on a piece of photographic film. The magnetic projector is analogous to the eyepiece of an optical microscope. The analogy is complete if an optical microscope is fitted with a projector lens that forms an image on a viewing screen. Microscopes so equipped can be used by a group of people rather than one at a time. And optical microscopes can also be equipped to project the image to photographic film.

If electrons bump into anything, their speed and direction is altered, and they are scattered. This is essentially what happens when the electrons pass through the specimen. However, bumping into the molecules that compose air would also scatter the electrons. If the electrons were scattered before they got to the specimen, a true image of the specimen would not be obtained. For that reason the electron source, the magnetic lenses, and the specimen must be in as good a vacuum as possible.

As was mentioned in Chapter 1, a glass tube that could be evacuated had to be made in order to discover that beams of electrons existed. A vacuum pump is part of every electron microscope installation, and the entire instrument must have the air pumped out of it before it can be used.

Specimens to be examined with the transmission-electron microscope must be fantastically thin. If they are too thick they might be destroyed by the electron beam. A usual thickness is about 100 A. Transmitted-light optical microscopes also give best results with thin specimens, but with the optical microscopes, specimens as thick as 25 $\mu$ (25,000 A) or more can be examined. Preparation procedures for specimens to be viewed in electron microscopes are usually quite elaborate and will be dealt with in more detail in Chapter 5.

There are many different kinds of electron microscopes made by many manufacturers to carry out certain kinds of observations. The transmission type, the first type to be made, is the most widely used, although others, notably the scanning-electron microscope are used in many special applications. However, there are certain things common to all and a transmission-electron microscope has an appearance distinctive enough so that if you have seen one electron microscope you know when you are looking at another one.

Most transmission electron microscopes look like a vertical cylinder sitting on top of a desk. The vertical cylinder contains the electron source (electron gun), the condenser, lens, specimen chamber, objective lens, and viewing screen. There are various protruding knobs which are controls for the components in the cylinder or tube (also called *column*). The "desk" has a console that houses more controls and various dials, gauges, and meters. The photographic film is usually in a chamber in the desk in front of the operator.

The electron gun is at the top of the cylinder. A heavy electrical cable leads into the gun assembly. The electron microscope is basically an electrical device which ultimately has to be "plugged in" to a source of electricity. Ordinary current supplied to homes

Scanning electron micrograph of a colony of bacteria growing on the eye of a fruit fly. The individual eye facets can be clearly seen (1,500X).

and factories has neither the voltage nor the stability for use in an electron microscope. Most houses in the United States are equipped with 110-volt or 120-volt service. For heavy appliances such as electric stoves 220-volt service is needed. To achieve the short wavelengths needed in electron microscopes, 80,000 volts or more of electricity are frequently needed. And the voltage supplied must be stable. That is, it must not vary. Voltage supplied from the power company usually varies by a few volts. Although the fuse or circuit box may be designed to deliver 120 volts, at any given time it might be 118 volts or 123. This does not make a great deal of difference in operating a toaster or an electric fan, but such variances can play havoc with an electron microscopist's work.

The heavy cable leading into the electron gun does not come from a wall plug, but from a set of transformers and voltage regulators. The transformers change the 220-volt house current to the 80,000 volts or whatever the operator needs, and the regulators keep it that way. The feed in and connection to the house current cannot be in the same room as the microscope. Magnetic fields generated by the current-breaker box, transformers, and other electrical equipment could adversely affect the magnetic fields in the microscope column. Electron microscopes are usually installed in a small, closetlike room with no windows. This serves to keep out dust and stray electromagnetic radiation. Such a room is also easily darkened to improve visibility of the fluorescent viewing screen.

The heart of the electron gun is a hairpin-shaped tungsten filament. The electron beam issues from the tip of the filament. The filament is surrounded by a cylindrical metal shield called the *Wehnelt cylinder.* The filament is not completely enclosed. A small hole in the Wehnelt cylinder, directly under the filament, serves as the passage for the electrons that stream from the filament tip. In terms of electrical potential the tungsten tip is the cathode or negative pole. The usual flow of electrons, including that in a wire, is from the negative pole to the positive pole or anode. The anode in the electron-gun assembly, which is located just below the Wehnelt cylinder, is a grooved metal cylinder with a hole for passage of electrons.

The Wehnelt cylinder helps to concentrate the electron beam. It acts somewhat as an electrostatic lens, reducing the electron-emitting area of the filament and resulting in a more concentrated electron beam. If a fluorescent screen was placed to receive the electrons just as they came out of the Wehnelt cylinder, an image of the filament tip would be formed.

The current of electrons that originates at the tungsten tip and flows through the holes in the Wehnelt cylinder and anode, and ultimately through the whole tube, is called the *beam current.*

The nature of the beam current can be controlled at the source by the operator. He (or she—many electron microscope technicians are women) can change the distance between the tip and the Wehnelt cylinder. He can also vary the electrical potential of the Wehnelt cylinder, which in effect changes its characteristics as a lens. The first operation is comparable to moving a magnifying glass closer or farther away from the object of view. The second is analogous to switching to another magnifying glass. The gun assembly can also be lifted and moved across, up, or down to achieve the proper alignment. Since tungsten filaments last only a few operating hours, the assembly is made so that it can be removed easily for filament replacement.

While the voltages used in electron microscopes are high, the actual current flowing through the tube is quite small, compared to the current in house wires. Voltage is potential or electrical pressure. It can be compared to the water pressure of a water supply. The current can be compared to the actual amount of water flowing through the pipe. The beam current would then be something like water under the great pressures from a fire engine pumper being pushed through a drinking straw. The water would come out in a fast, energy-packed flow. The electron beam is very fast and contains a great deal of energy.

High voltages are desirable in electron microscopes. A general rule is that the higher the voltage, the greater the frequency and, therefore, the smaller the wavelength of the electron beam. The smaller the wavelength, the smaller are the objects that can be detected and resolved. With sufficiently high voltage, wavelengths of 0.05 A can be achieved. However, lens limitations usually limit resolving power to 1 or 2 A at the best. Occasionally resolutions of less than 1 A are achieved.

The condenser lens concentrates the beam current and focuses it in the direction of the specimen. The condenser can be controlled to change the brightness of the final image. Most instruments have more than one condenser lens. The number of condenser lenses is determined by the manufacturer and the particular applications for which the instrument is designed. Some lower-priced instruments, designed mainly for demonstration use in schools and colleges, do not have condensers. In the same way, less expensive light microscopes used in schools frequently do not have condensers.

The condenser can be manipulated to increase or decrease the illuminated portion of the image or to spotlight specific areas. The controls are usually located on the tube at the position of the condensers.

The material to be observed is placed in a holder which is put into a specimen chamber. The design of the chamber varies widely, but it is usually between the condenser and the objective. There is generally a door in the tube for access to the chamber. The chamber is equipped with air locks; vacuum is lost only when the door is opened. Some higher-priced instruments have rotating specimen-holder turrets so that several holders can be available for viewing without breaking the vacuum in the chamber each time a specimen change is desired. This is a great time-saver. In some instruments access to the specimen chamber is through a hole in the objective-lens assembly.

The specimen holder is a brass cylinder or cone with a seat to receive the specimen. Electron microscope specimens are mounted not on glass slides but on thin circular metal meshes called grids. The grid is placed in the seat in the holder, and the holder is then placed in the specimen chamber. The grid and holder will be described more fully in the chapter on techniques. The proper alignment of the holder in the specimen chamber is achieved by controls on the tube and console.

The objective lenses are the ultimate determiners of the instrument's resolving power. Magnetic lenses have been greatly improved since the 1920s, but they are still basically of the same design as those made by Busch, Knoll, and Ruska. And they still have many of the same problems.

An electromagnetic lens is essentially a tightly wound coil of wire in a doughnut-shaped soft iron shell. When a direct current of electricity is sent through the windings, a magnetic field results. In some lenses two pieces of iron, collectively called the *pole piece*, are inserted against the inner surface of the doughnut. The two parts of the pole piece are separated by a piece of nonmagnetic metal. The pole piece serves to concentrate the magnetic lines of force in the small space between the components of the pole piece. This helps to keep the electrons flowing in the proper axis and to bring the beam current to the proper point of focus.

An electromagnetic lens is more flexible in use than a glass lens. Changes in focal length, magnification, and image illumination are brought about by changing the intensity of the current in the windings. It is not necessary to change or move the lenses. The lens controls are usually located on the console.

Electromagnetic lenses have many of the same problems as glass lenses. The major problems are spherical and chromatic aberration and astigmatism.

Spherical aberration in electron microscopes is similar to that in glass lenses. The point of focus in the center of the lens is different from that in the outer parts. It is impossible to have the entire image in focus at one time. To reduce aberration the hole through which the electrons pass must be very small. However, decreasing the size of the hole tends to reduce resolving power. In any given lens there is a size of hole that combines the greatest resolving power with the least spherical aberration. Much of the continuing research in lens design is devoted to increasing the resolving power while minimizing aberration.

A vacuum evaporator apparatus.

The size of the lens hole or aperture can be adjusted in an electron microscope. Smaller apertures tend to give better contrast, that is, the blacks and grays stand out more clearly. The choice of aperture is largely a matter of experience.

Since the image produced by an electron microscope is black and white, chromatic aberration does not involve halos of color as in optical microscopes. It results in blurring of the image as electrons come to focus at different distances from the lens and in that respect is similar to chromatic aberration in a light microscope. It is caused by changes in the energy of the electrons. This can result from voltage fluctuations or attempts to view a specimen that is too thick. The remedies are to prepare thin specimens and to maintain good voltage regulation.

Astigmatism is blurring, due to lenses that are out of alignment. It can also be caused by dust and dirt in the electron-beam path. To prevent astigmatism, the instrument must be kept clean and the lenses held in proper alignment—which is not as easy as it might sound. Static electricity builds up in electron microscopes and attracts dust particles. Magnetic lenses are never quite perfect, and the slightest imperfections cause astigmatism. Since all astigmatism cannot be eliminated in the manufacture of the lens, most instruments are equipped with a *stigmator*, which compensates for lens imperfections. The controls for the stigmator are located on the tube. A widely used type of stigmator consists of a circle of electromagnets mounted around the objective lens. The magnets are adjusted to balance out magnetic deviations in the objective lenses.

Electron technicians usually adjust for astigmatism before examining a specimen. The controls are manipulated until there is a clear, bright spot of light on the viewing screen. The spot is then enlarged, and if it does not enlarge evenly, astigmatism is present. Further adjustments are made with the stigmator until the desired light-spot characteristics are obtained.

The projector lens, near the base of the tube, enlarges the image formed by the objective lenses and projects the image on the viewing screen or photographic film. The system usually consists of at least two lenses. Use of more than one lens tends to reduce distortion, as the distortions produced by one lens tend to cancel out those of the other lens.

Many instruments are equipped with variously sized projection-lens apertures, which enable the operator to project any part of the image. These are useful in certain types of experiments. For example, the way in which a specimen diffracts (scatters) electrons may give information on the specimen's chemical composition. The apertures can be used to obtain sharper definitions of diffraction patterns. Some projection systems have a diffraction lens that further defines the diffraction patterns and reduces distortion.

The viewing screen is usually housed in a windowed enclosure in front of the operator. There are many variations in the number and shape of the windows. There is usually a low-power (about 10x) binocular microscope mounted on a hinge for close examination of the image. In some instruments a "mask" in the viewing enclosure blocks out the bright central part of the image to clarify diffraction patterns on the edge.

The camera is usually located under the viewing-screen chamber. Its structure varies from one instrument to the next, but essentially it is a box that holds roll or cut film. Each piece or section of film is fed into place automatically. The film chamber must be in vacuum, and it is equipped with air locks so that film can be reloaded without breaking the vacuum in the entire instrument.

In most instruments the viewing screen is the shutter. By means of a lever, the viewing screen is flipped, pushed, or pulled out of the way so the electron beam can fall on the film. A fine-grain film is usually used.

Electron microscopes may also be hooked up to television monitors, and the images can be recorded on video tape.

The components discussed up to this point are basic to all transmission electron microscopes. There are hundreds of attachments, accessories, and special installations designed for a variety of purposes. A few of the more common ones can be mentioned here.

Devices for tilting and altering the axis of microscope components achieve various effects. A stereo effect can be obtained by tilting the specimen grid. The grid is tilted slightly, and a picture is taken. The grid is then tilted in another direction, and a second picture is taken. The two pictures are mounted in a stereo viewer for a three-dimensional effect.

Accessories that help to keep the instrument clean are understandably popular among electron microscope technicians. Oil in the instrument is a problem. The machinery of the vacuum pumps that evacuate the electron microscope is swimming in oil, some of which escapes into the microscope. The amount is so small that it cannot be seen, but it seriously hampers the instrument's operation. The high-energy electrons decompose the oil, and the remains build up on the specimen, in the holes through which the electron beam passes, and on the inner wall of the tube. Astigmatism is increased, and if the contamination becomes too heavy, there is no image at all.

One decontamination device is a circular object or disc that is placed between the specimen and the objective. It is supercooled with liquid nitrogen. The oil vapor and the products of decomposition condense on this cold disc. It is used when the microscope is not in operation. Of course, the disc itself has to be cleaned from time to time, but that is much easier and takes less time than cleaning the whole instrument. Some microscopes are equipped with attachments that cool the tube. This also helps to keep the oil vapor out.

Goniometers facilitate tilting of the specimen. In advanced models one knob rotates and tilts the specimen to the orientation desired.

There are special types of specimen holders for examination of wet specimens and gases. There are attachments for cooling or heating the specimens. Since the entire system must be in vacuum, living specimens usually cannot be examined. Additionally, water must usually be removed from specimens. Water vapor in the vacuum of an electron microscope column would make the instrument useless. But dehydrated specimens are not "natural." If specimens could be viewed with their water content still present, the image would be more like the living thing.

Some manufacturers have made elaborate specimen holders with which wet specimens can be viewed. One of these holders is a small cylinder. When it is in place in the specimen chamber, a small tube extends from it to the outside of the microscope. The openings are covered by a fine metal screen and sealed with a thin plastic film. The film keeps the water in the specimen from vaporizing into the microscopic column and is strong enough to stand up under the difference in pressure between the inside and outside of the microscope. Electrons can get through the plastic film to the specimen, but very high voltages are needed. The

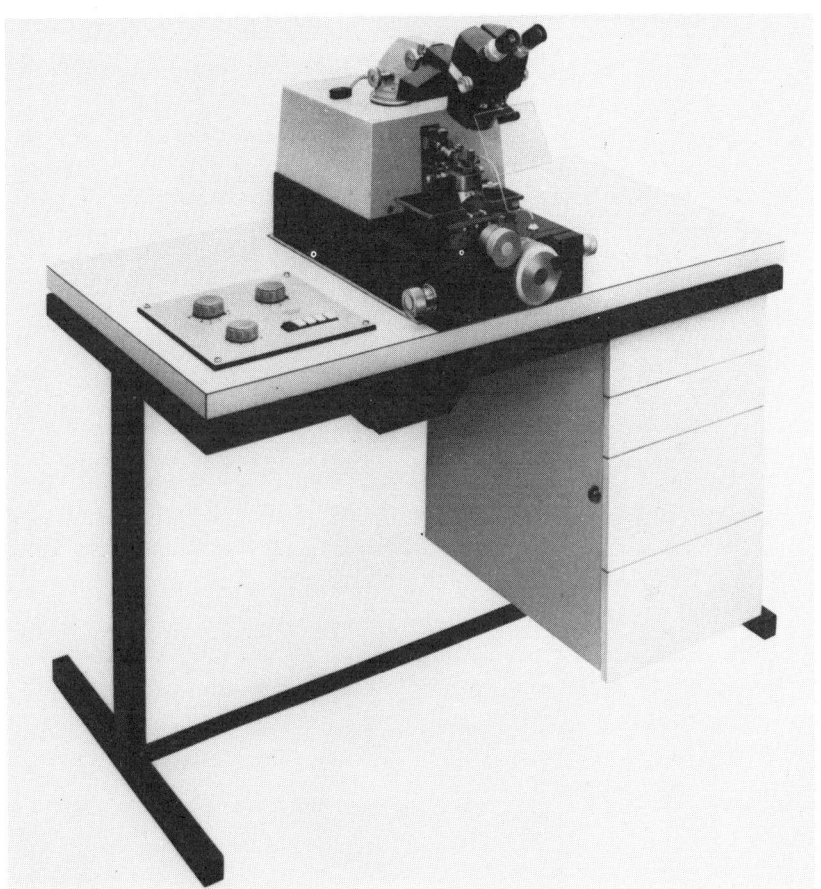

An ultramicrotome.

resulting image is weak, and a long exposure time is required to get a picture. With this technique live bacteria have been seen, although not very clearly. With the development of higher-voltage instruments the results obtained with this technique may be improved.

The number and nature of attachments on an electron microscope are determined by its primary use and by the amount of money the laboratory has to spend. A question frequently asked by visitors to electron microscope laboratories is: How much did it cost? It is as difficult to answer as the question: How much does an automobile cost?

Electron microscopes are very expensive, but individual prices depend on many factors. Each microscope must be assembled by hand, and the precision required in every manufacturing step adds greatly to the cost. In the 1970s the minimum price would be about $25,000. Elaborate installations may require expenditures of millions of dollars. There are electron microscopes that sell for less than $25,000, but these are mainly teaching instruments which are not suitable for research. These relatively low-priced instruments enable small colleges and universities with limited funds to provide some electron microscope experience for their science majors.

The expense of electron microscope research only begins with the purchase of the instrument. Installation, adequate wiring, floor space, and other requirements add to the cost.

The most expensive and elaborate electron microscope is useless without a technician to operate the instrument and, more importantly, to prepare the specimens, an elaborate, complicated procedure. Well-trained technicians constitute a continuing expense in any electron microscope facility.

# 5
# Techniques

Electron microscopes were developed before anyone knew anything about how to prepare the specimens. One thing was immediately evident: Whatever was examined in the instrument had to be very thin. By the 1930s, when electron microscopes were introduced, techniques of preparing specimens for the light microscope had become quite sophisticated. Cutting machines called microtomes could cut sections of tissue (groups of cells) as thin as 5 microns. A wide variety of techniques was available to laboratory technicians who prepared cells, tissues, and other materials for microscopic examination. A typical light-microscope procedure involves embedding the tissue in a block of paraffin. This is to hold the tissue for placement in the microtome. The thin sections cut by the microtome are mounted on glass slides. Stains are added to bring out certain features in the specimen. A cover slip placed on a drop of adhesive liquid over the specimen makes the preparation more or less permanent.

Histological techniques for light microscopes (*histology* is the study of tissues) were advanced, but it was apparent that few, if any, were applicable to electron microscopy. Tissue sections for electron microscopes had to be about 0.1 $\mu$ in thickness, and when electron microscopes were first introduced, no microtome could do that. Most of the stains used for light microscopy were useless in electron microscopes. Many scientists thought that the electron microscope would never be of much use because speci-

Scanning electronmicrographs of a radiolaria, a single-celled organism (330X).

mens of sufficient thinness were impossible to obtain. Many biologists were convinced that the specimens would be destroyed as soon as the electron beam passed through them.

In the thirty or so years that electron microscopes have been in use, many techniques have been developed. Microtomes capable of cutting 0.1 $\mu$ sections were soon made available. Various techniques and instruments have enabled scientists to obtain information about many materials, both living and nonliving. New techniques are being developed all the time, but there are still many problems. Electron microscope techniques require skill, patience, and ingenuity. The procedures are laborious and painstaking, and every technician has his or her own bag of tricks. There are many procedures, but two goals are common to all. The specimen must be thin enough to allow electrons to pass through, and the specimen must scatter the electrons in the right places to produce the image that will give scientists the information they seek.

The various techniques can be classified into three general categories: thin sectioning, replicas, and direct spraying or dropping of liquids and suspensions. No classification scheme can include all the possibilities, but most techniques involve these three operations or variations of them.

Materials for examination in light microscopes are usually placed on glass slides. Specimens for electron microscope examination are placed on circular metallic, electroplated grids. The grids are made of a very fine metallic mesh. They are like Lilliputian window screens. A close look at a window screen will reveal a pattern of many tiny squares. A magnified look at a specimen grid will also reveal many tiny squares, but the squares are much tinier than those of a window screen.

The grids can be bought in a wide variety of sizes and fineness of mesh. A commonly used type has 200 squares per linear inch. The choice of type depends on the nature of the material to be examined.

The grids are made in the mesh pattern so the electrons can get through the openings. The electrons will not pass through the metal-wire part of the grid. It would seem that an object so small that an electron microscope is needed to see it would fall through the squares, even if they are small enough to fit two hundred to the linear inch. That is a correct assumption. Before a specimen can be mounted, the grid must be coated with a very thin film of a plasticlike material, strong enough to provide support for the specimen, but thin enough to allow the passage of electrons. So one of the continuing tasks of electron microscope technicians is putting film on grids. Some of the ways of doing this will be discussed here.

The choice of supporting film material is determined to some extent by the grid mesh. A 100-per-linear-inch grid will need a thicker, stronger film than a 200-per-inch grid. The supporting wires are farther apart on the 100-per-inch grid.

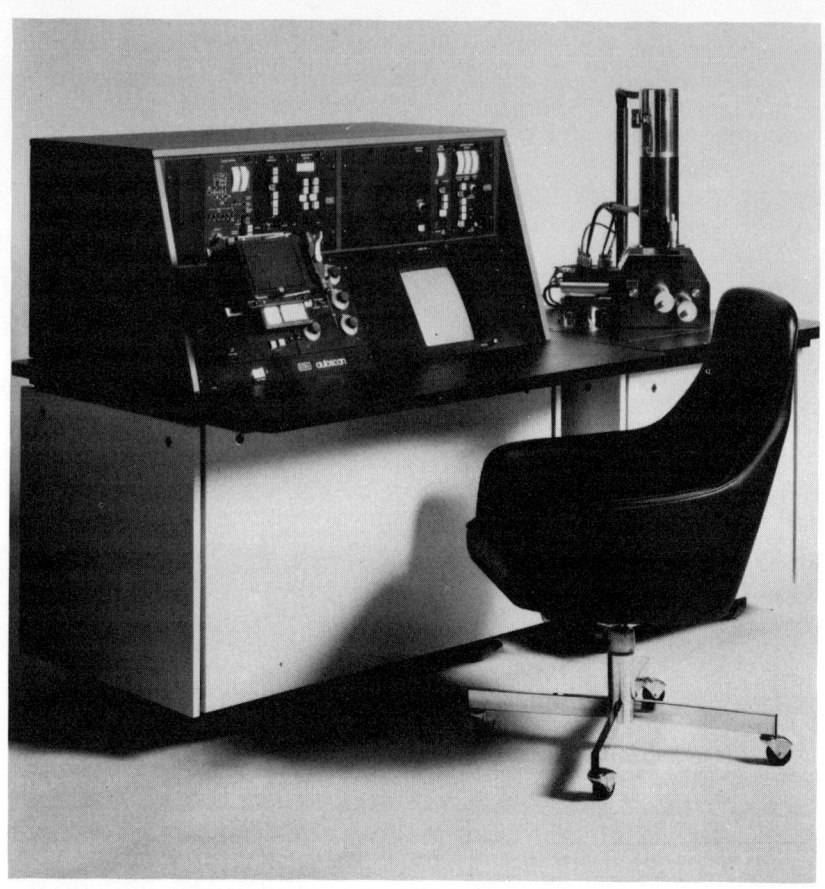

The AMR-1000. A compact, advanced model of a high resolution scanning electron microscope. The column is much smaller than that of most transmission electron microscopes.

A popular method is the spread-film technique. A drop or two of liquid plastic is placed on the surface of some water in a container such as a small bowl. This causes a thin layer of supporting film to spread out and float on the water. The grids are brought up through the water against the film, which will adhere to the grids. The operation requires the right touch and, as with any skill, the results improve with practice.

Each technician has his or her own technique, but the following is a typical procedure: A shallow circular dish is filled to the brim with water. A *grid spade*, which is something like a pancake flipper or spatula, is placed on the bottom of the dish. Then with a forceps, grids are carefully placed on the immersed spade. Grids are always handled with forceps, never with fingers. The spade handle is bent so that the blade can rest flatly on the bottom and the handle will still be accessible to the technician. The surface of the water is swept clean with a metal bar.

A frequently used material is a solution of a plastic called *collodion*. The collodion is dissolved in a volatile solvent such as amyl acetate, usually in a concentration of 3 parts of collodion to 97 parts of the solvent. A glass rod is dipped into the collodion solution. The drop that adheres to the glass rod is then touched lightly to the water surface. If done right, the collodion will spread out in a thin floating film. The spade is then slowly brought up against the film and lifted out of the water, leaving a thin layer of collodion covering the grids. Excess water is blotted up, and the grids are allowed to dry thoroughly. The grids are picked up with a sharp pair of forceps. The knife-edge of the forceps cuts the grids away from the sheet of plastic that forms on the spade. The filmed grids are usually kept on glass slides and stored in a way to protect them from dust and dirt. Other plastics similar to collodion, sold under a variety of trade names, are also used for filming.

Plastic films are not suitable for all procedures. They sometimes break under the impact of high-intensity beam currents, especially when the particles on the film are heavy. Fairly strong films can be made by *vacuum evaporating* certain substances in a way so that they coat the grid surface. Sometimes the coats are deposited over grids that already have a collodion coat.

The vacuum evaporator is a piece of equipment found in every electron microscope laboratory. It is used not only to film grids, but also to coat specimens with various substances that increase the specimen's "seeability" in the electron microscope.

The vacuum evaporator is an enclosed glass dome connected to a vacuum pump. Electric leads pass into the dome. The dome rests on a base plate that can be removed so that the work can be put aside. A tungsten filament is attached to the leads, and when a current is sent through, the filament is heated to high temperatures.

The grids to be coated are put into the evaporator. A small amount of the substance to be evaporated and deposited on the grids is placed on the filament. The evaporator is closed, sealed, and pumped out. When the filament heats up, the substance on it is evaporated or vaporized. Since this is taking place in a vacuum,

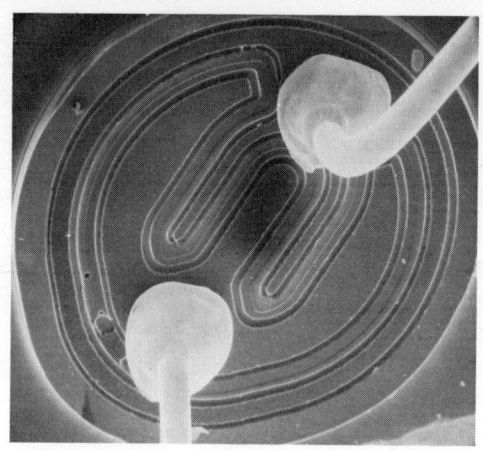

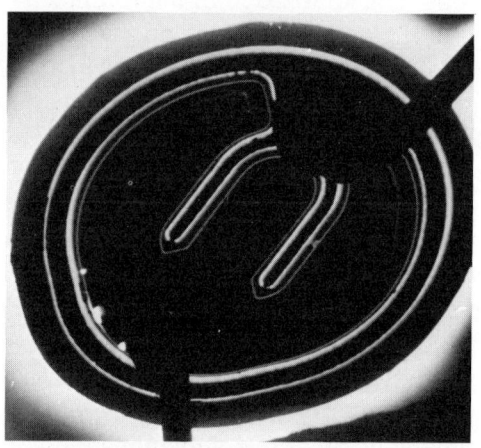

A scanning electron micrograph of a transistor (50X).

the particles spread out evenly and deposit in an even coating on the grid.

The filaments can be made of metals other than tungsten, and the variety of filament-tip shapes is endless. The shape of the tip can make a difference in the results. Many technicians fold their own.

Some substances, desirable for grid films, do not spread on water or vaporize. A favorite technique with these substances is to form the film on a glass slide first. This is done by dipping the slide into the liquid substance, which forms a film when it dries. The film is then sliced so that it will slip off and float when the slide is carefully immersed in a bowl of water. Grids can then be filmed as in the spread-film technique.

The simplest of the three general categories of specimen preparation is placing a small drop of liquid material on the grid. Sometimes suspensions are sprayed on the grid. This tends to give an even coating and is useful for taking virus or bacteria counts. In this procedure, the specimens, such as viruses or bacteria, are suspended in the liquid. The liquid is frequently water, but it can be any number of liquid substances. Of course, all liquid must be evaporated before the grid can be put into the electron microscope. After the specimen dries, it it treated with various chemicals. The chemicals are called *stains* but these are not stains in the same sense that the dyes used in light microscope work are stains. Electron microscope stains are used to increase the contrast of the image and to make certain substances do the right things to the electrons so that a meaningful image will result. There are two general types of stains. Negative stains change the electron-scattering characteristics of the background in a way that improves visibility of the specimen. Positive stains act directly on the specimen.

Dropping and spraying techniques are relatively simple but many problems can, and frequently do, occur. The film can break, and most technicians examine the grid for film breakage with a low-power light microscope at several points during the procedure. Continuing to carry out the preparation when the film is broken is like putting water in the bathtub when the plug is out.

The grid has to be washed off with a thin stream of distilled water between applications of chemicals. If this is not done, the various chemicals might react and form unwanted substances on the grid. The washing can sometimes rinse the specimen materials off the grid. Some substances tend to stick to the grid, and others do not. When working with substances that are likely to wash off, the technicians add chemicals to overcome the problem, though the kinds, order, and quantity of chemicals used is again a matter of experience.

The surface of objects that are too thick for transmission electron microscopes can be studied by making thin *replicas* of the surface on a thin film. As the word "replica" implies, the technique involves making a mold or impression of the specimen surface. The replica must then be stripped off the specimen and placed on a grid. This procedure is widely used in studying metals and is also useful in some biological procedures.

The easiest replicas to prepare are negative replicas, so called because they are the reverse of the object. This is like pressing a piece of clay against a coin. The image of the coin in the clay is negative. Where there is a bump in the coin, there is a depression in the clay.

A solution of a plastic material such as collodion is spread evenly over the surface of the material to be examined and allowed to dry. As it dries, the replica is formed. This is somewhat like spreading plaster over a shallow mold. There are several ways of transferring the replica to a grid. One commonly used method will be considered here. The dried plastic-film replica is cut through at two ends. Then the technician breathes on it. The moisture in the breath enters the film and makes it easier to separate it from the object. Then a number of unfilmed grids are placed on the plastic-film replica. Paper discs are placed over the grids, and a piece of adhesive tape is pressed down over the grids. The paper discs keep the tape glue from getting on the grids. The tape is then gently lifted, the replica strips off the object carrying the grids and adheres flatly and snugly to the grids. The grids are then removed with sharp·forceps. The replica, now on the grids, is then treated with chemicals to bring out whatever features of the specimen are desired.

Replicas can also be made by vacuum evaporation. The vaporized materials from a coating on the object and the object is stripped or dissolved off the object.

Positive replicas require another step. A negative replica is made as described above or by other methods, and a replica is made of the replica. The second replica is positive. That is, it has bumps and ridges where the original object has bumps and ridges. There is a wide choice of methods and materials. However, the plastic used to make the negative cannot be used to make the positive. In this procedure the negative can be, and usually is, made of a thicker film than collodion. Since the negative will not be examined in the microscope, it is usually made of a thicker plastic that is easy to handle. The plastic for the positive is then spread over the negative, allowed to dry, stripped off, and mounted on

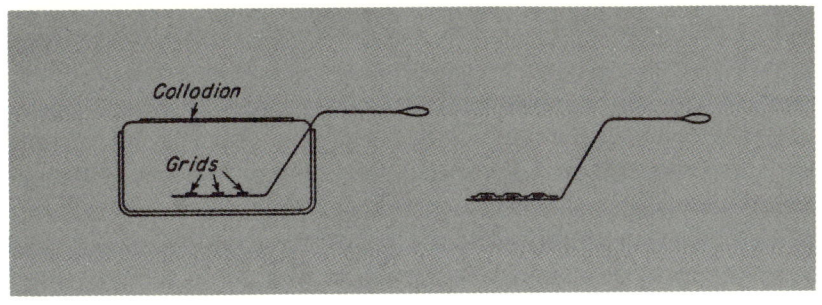

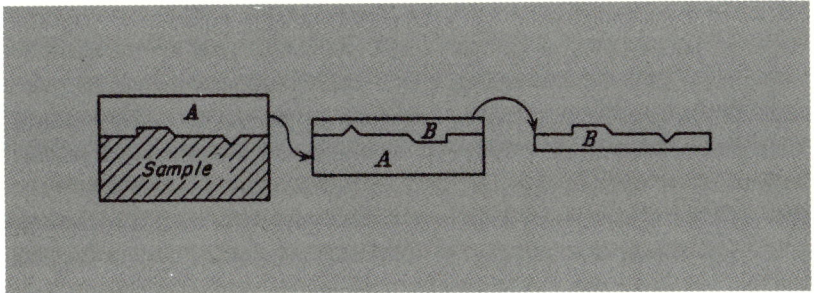

Top: Settings for filming grids. The diagram shows method for lifting grids up against collodion film.
Bottom: The diagram shows the molds used for positive replica techniques.

grids in a variety of ways, such as stripping, floating, or dissolving.

Thin-sectioning is a technique inherited from light-microscope technology. However, the legacy has been much changed and electron-microscope sectioning techniques are far more exacting and involved than light-microscope techniques. As noted earlier, sections for electron microscopy must be unbelievably thin. Thin-sectioning is most useful in studying the structure of cells and tissues.

The instrument that is used to cut thin sections is called a microtome. The basic parts of a microtome are a knife, something

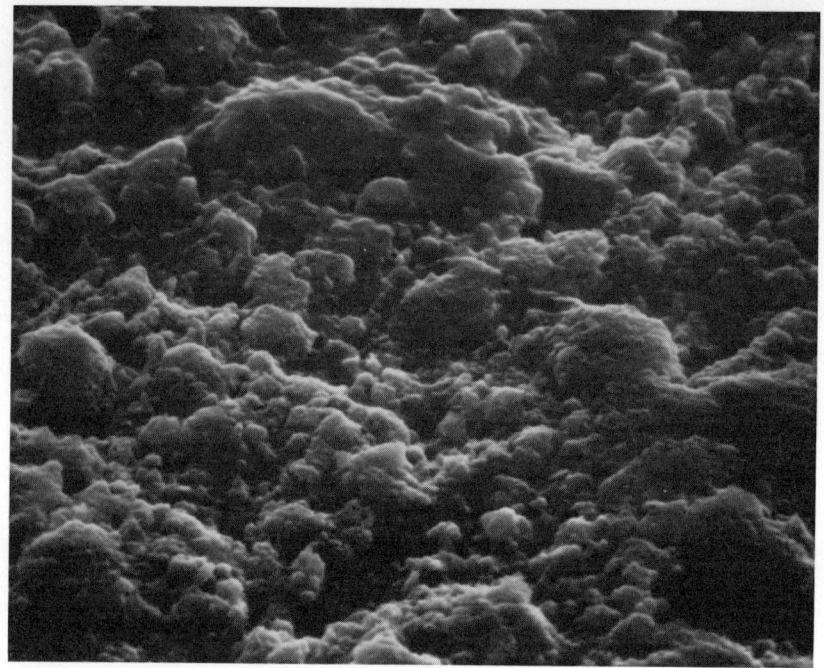
Scanning electronmicrograph of high quality coated paper used in "slick" magazines (2,500X).

to hold the specimen, and something to move the specimen to the knife as a section is cut. The most sophisticated microtomes are rotary types. In rotary microtomes the knife is stationary, and the mechanism that feeds the specimen to the knife is controlled by a wheel on the instrument. It can be run by a motor or operated by hand.

Microtomes for electron microscopy are called *ultramicrotomes.* As the name implies, the feed mechanisms are very exact and can advance the specimen 100 A at a time or less. Ultramicrotomes are equipped with low-power stereo microscopes. The sections are usually so small and thin that the technician cannot see what he is doing without the microscope.

When microtomes were first tried for cutting sections thin enough for electron microscopy, it was found that paraffin would not support the thin sections. They crumbled on handling. It was obvious that a much harder embedding substance was needed. However, the steel knives used in light-microscopy microtomes could not effectively cut the harder material. After much experi-

Scanning electronmicrograph of the surface of the tongue of a fly (2,400X).

menting it was determined that the best knives were glass and diamond. The edge of a broken sheet of glass is very sharp. Most technicians made their own glass knives. Knife-making machines break sheets of glass at the best angle and edge depressions to achieve the sharpest edge. Glass knives do not last very long but glass is cheap, and the blades can be shifted as the part of the knife used for cutting becomes dull. Diamond knives last considerably longer but are considerably more expensive than glass knives. Many technicians prefer glass over diamond, maintaining that glass gives them better results.

The most frequently used embedding materials are epoxy resins. The material is quite hard, transparent enough so that the embedded specimen can be seen, and specimens embedded in epoxy resins can be kept indefinitely without refrigeration.

The embedding process is a long, drawn-out affair. The beginning steps are similar to light-microscope histological techniques. A piece of tissue, such as liver, muscle, or any other animal or plant tissue, is cut up into small pieces. The tissue is then put into a

A rose-like pattern is seen in this 2,000X scanning electronmicrograph of the zinc coating on galvanized iron.

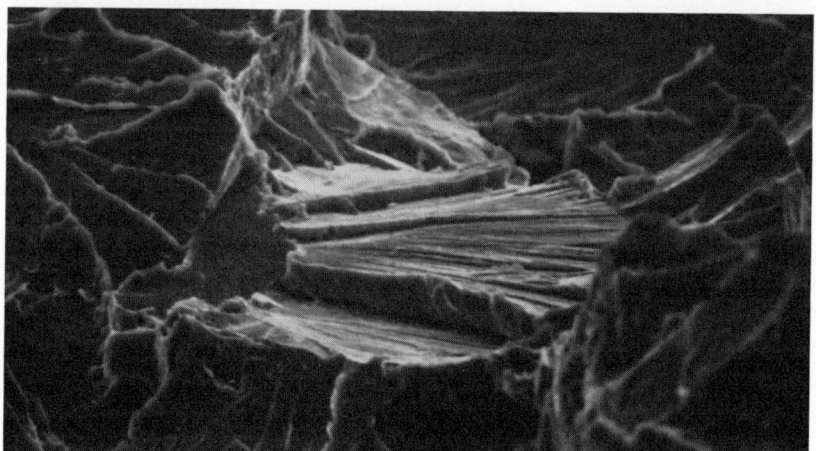
The broken surface of stainless steel suggests the interior of a cave (2,400X).

fixative solution, which acts as a preservative. Some fixatives also make the tissue more receptive to stains and other chemicals that may be used later. The tissue is then dehydrated. The dehydration process involves placing the tissue in successively higher solutions of ethyl alcohol and water. Each step removes additional water. The alcohol series might start with 35.0 percent and advance to 100 percent alcohol. The steps, or "jumps," depends on many factors. If the tissue is particularly fragile, the increase might be in 10 percent or even 5 percent steps. The steps are usually

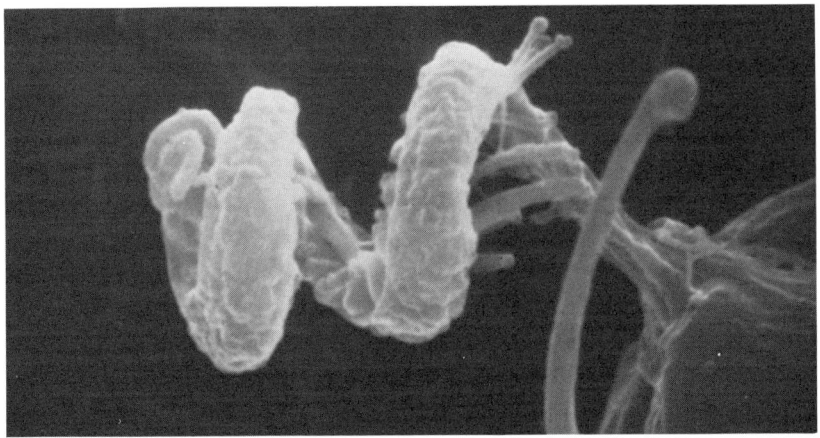

Scanning electronmicrograph of a stinging cell of a hydra (1,500X).

Scanning electron micrograph of human hair (1,000X).

longer. A typical series might be 35–60–70–80–95–100. The specimen is then gradually infiltrated with the resins so that the entire specimen will be supported. The final embedding is done in a small metal, bullet-shaped capsule. The specimen embedded in the hard epoxy material is called the block. The block is trimmed and placed in the microtome. A small trough of water is situated close to the knife. The sections are floated as they are cut and then are picked up on grids. Staining and other chemical treatment are then carried out as needed.

# Picture Essay:
# ELECTRON MICROSCOPE TECHNIQUES

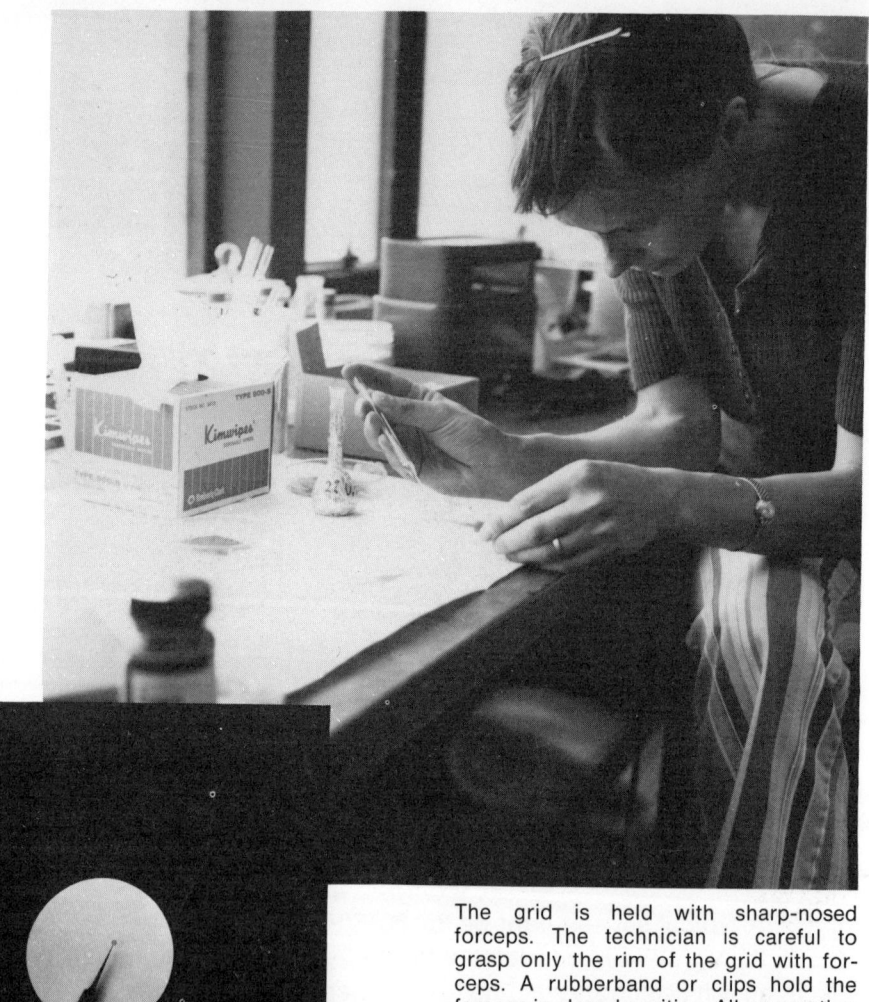

The grid is held with sharp-nosed forceps. The technician is careful to grasp only the rim of the grid with forceps. A rubberband or clips hold the forceps in closed position. All preparation operations are carried out with the grid held by the locked forceps.

The technician is placing a drop of virus suspension on the grid. Some technicians prefer to hold the grid during the dropping. Others prefer to place the forceps on a piece of filter paper on the laboratory table.

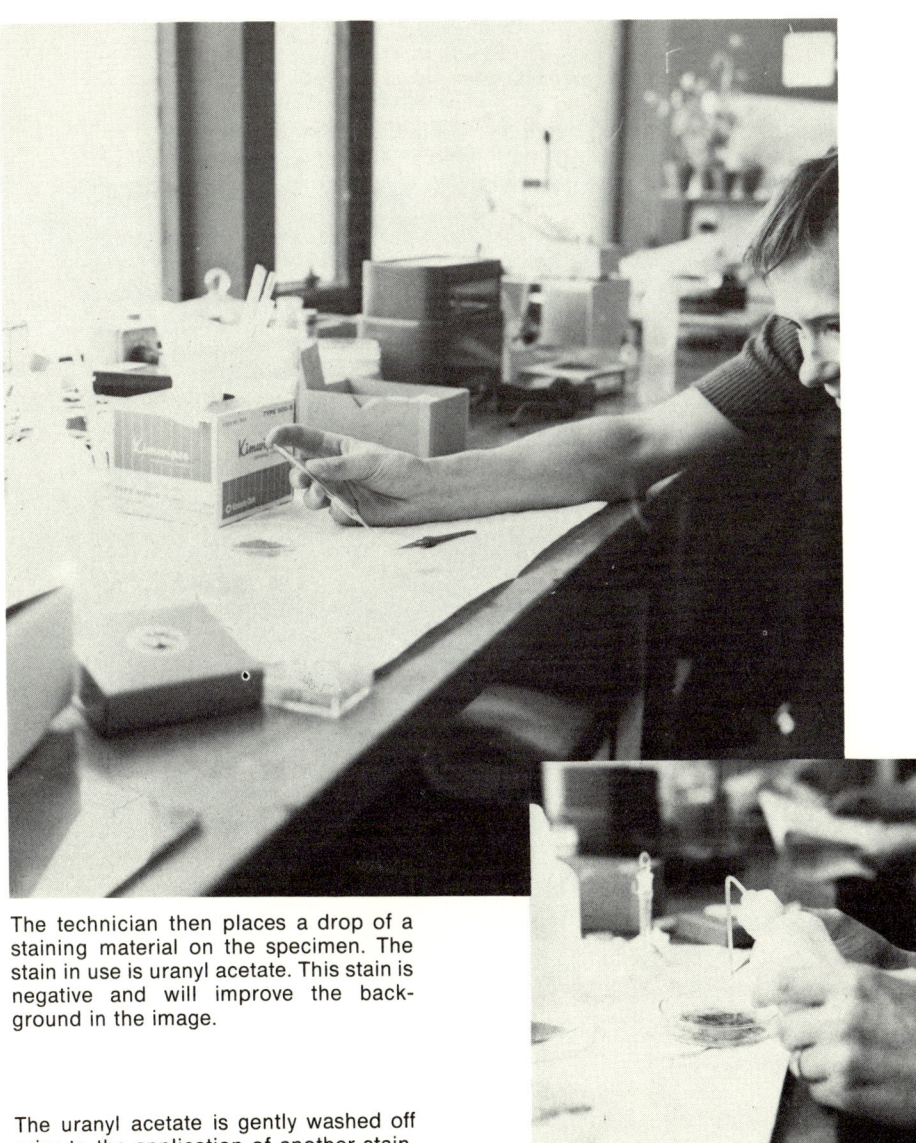

The technician then places a drop of a staining material on the specimen. The stain in use is uranyl acetate. This stain is negative and will improve the background in the image.

The uranyl acetate is gently washed off prior to the application of another stain. The step must be carried out very carefully to avoid washing the specimen off the grid.

The next stain is a lead compound. It is positive and is used to improve contrast. The grid is placed in a petri dish in which there are pellets of sodium hydroxide. The sodium hydroxide is to absorb carbon dioxide to prevent the formation of lead carbonate on the specimen. The lead carbonate would show up as black spots in the image. The specimen could be shadow-cast if desired.

Specimen holders or cartridges are kept in boxes to protect them from dust and dirt. The larger objects at the left are modifications for a rotary specimen holder.

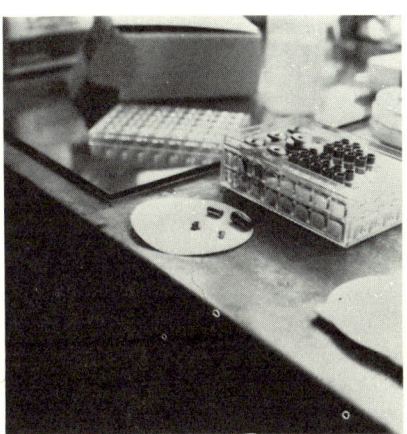

The specimen holder must be disassembled before the grid can be inserted. The object in the center of the filter paper is the part that receives the grid.

The grid is carefully placed in its "seat."

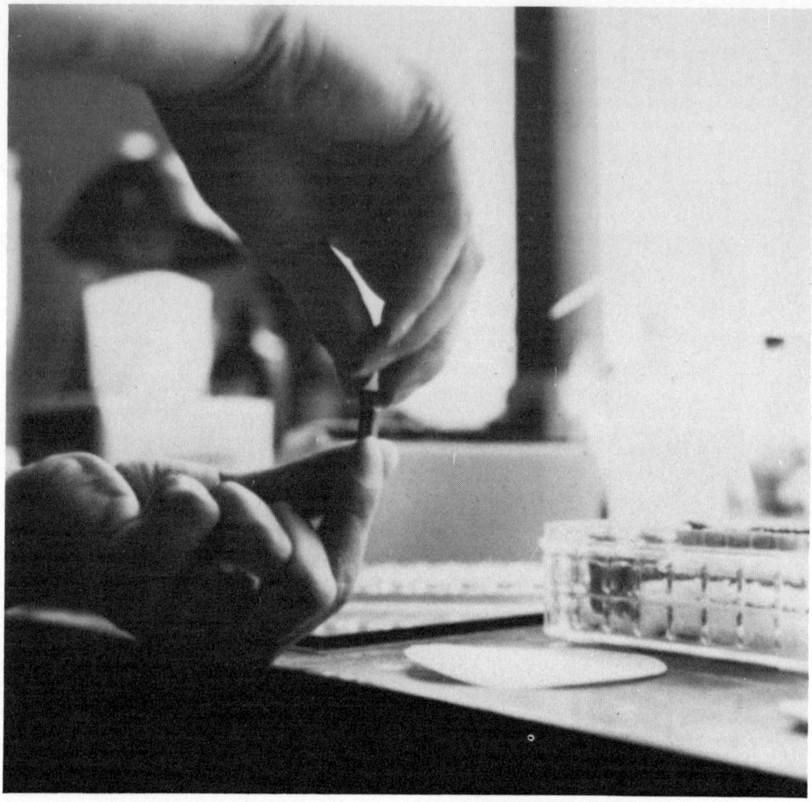

The specimen holder is reassembled.

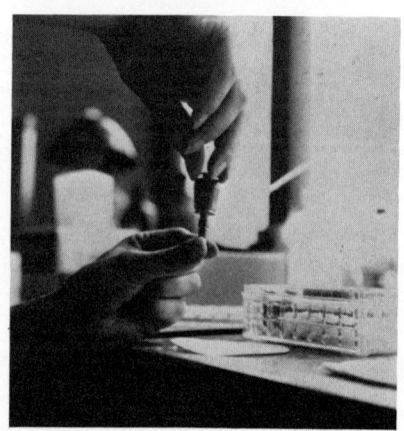

The revolving turret adapter is screwed into place.

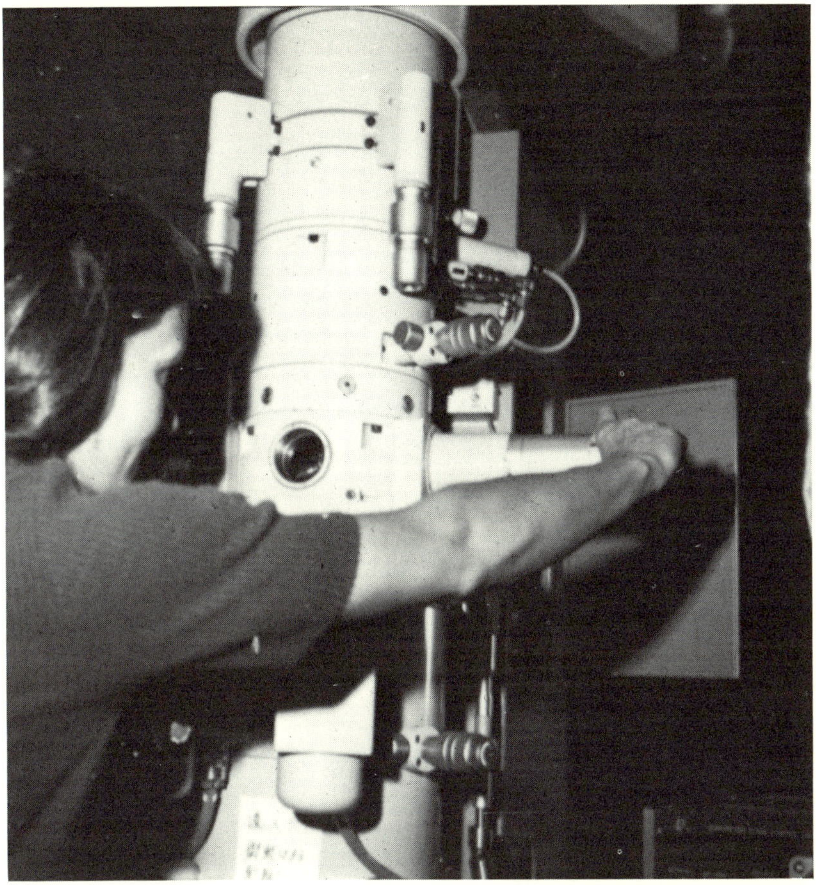

Before opening the specimen chamber door, the technician closes the air lock. This allows the vacuum to remain in the column. Only the specimen chamber will have to be re-evacuated.

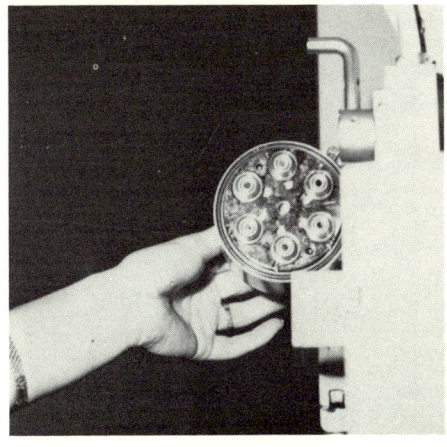

The specimen turret is in the door. There are six specimen holders in place. A mechanism in the specimen chamber lifts the holder out and maneuvers it into proper position. The "porthole" allows the technician to look into the specimen chamber.

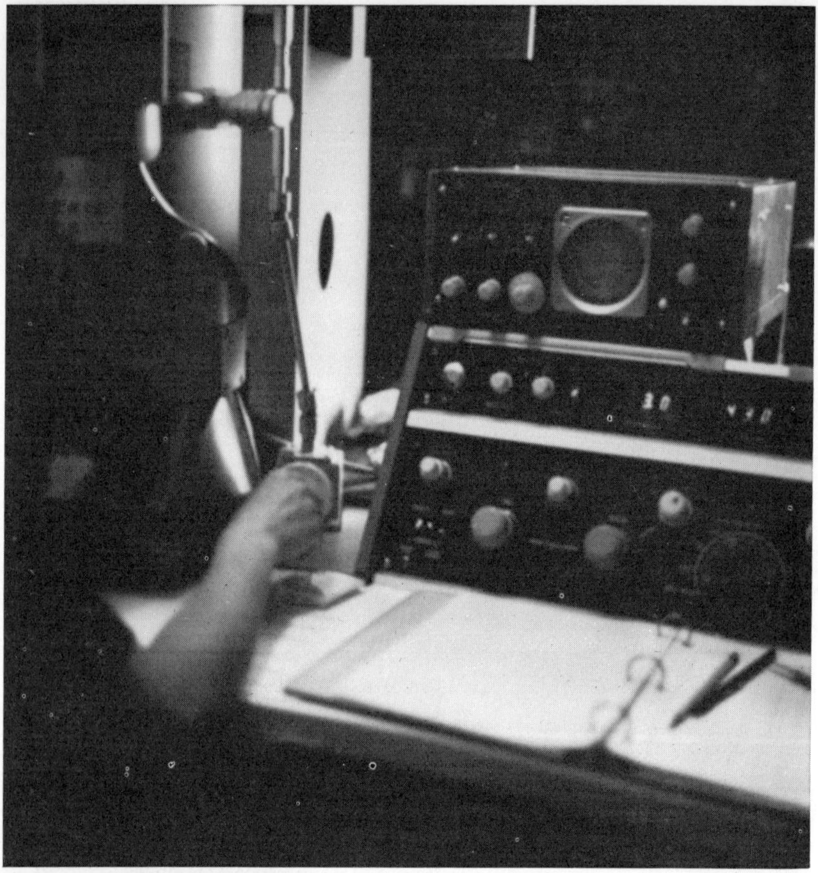

The electron microscope is in operation and the technician is viewing the image on the fluorescent screen. Controls are to the right of the technician. The one on the top with the oscilloscope screen is for various preliminary testing. Other controls are for magnification, brightness, focusing. There is also a digital gauge that indicates how much film is left in the camera.

The film boxes are in a chamber in the "desk." The door cannot be opened while the film chamber is in vacuum. In order to open the door, the chamber must be sealed off with an air lock and the vacuum broken. The door then opens automatically.

For thin sectioning, the specimen must be imbedded in a "block" of epoxy resin. The technician is holding a block. The specimen is in the end of the block that is pointing up.

An ultra microtome placed on a work table. The instrument can be operated by an electric motor or manually, with the handled control wheel at the right. The instrument to the left is a glass knife maker.

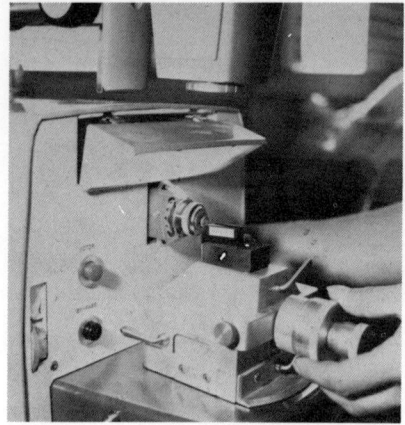

The block is placed in the holder, which is the round, metallic structure under the lamp. The knife is just in front of the oblong black trough. The trough is filled with water. The holder moves up and down, and the cutting is on the down stroke. The sections fall into the trough and float. The technician adjusts the microtome for thickness of cut, speed, and rate of advance.

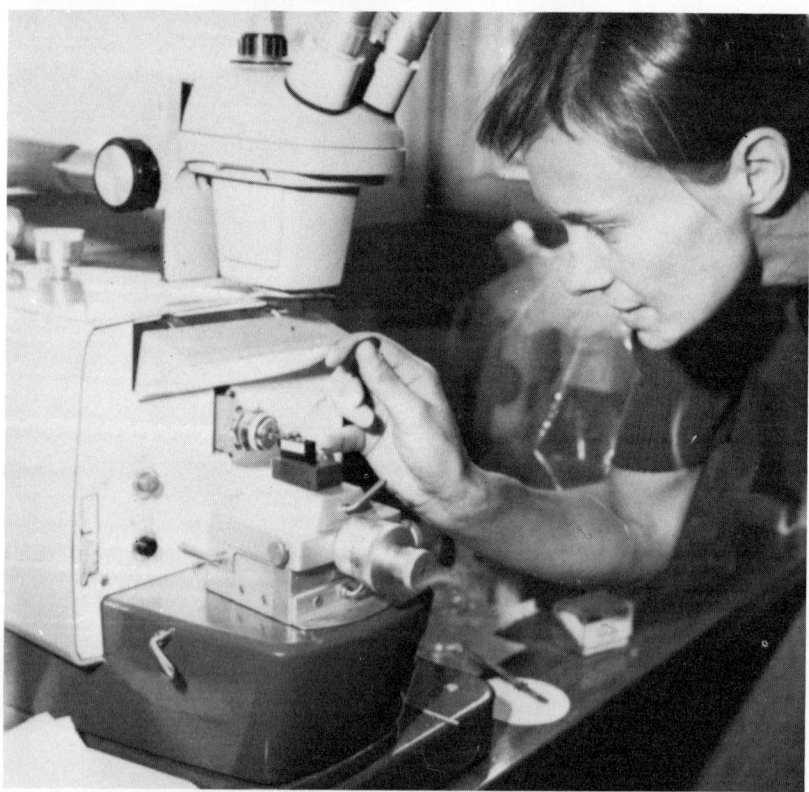

As the ultra microtome operates, the technician sometimes has to gently tease the sections into the trough.

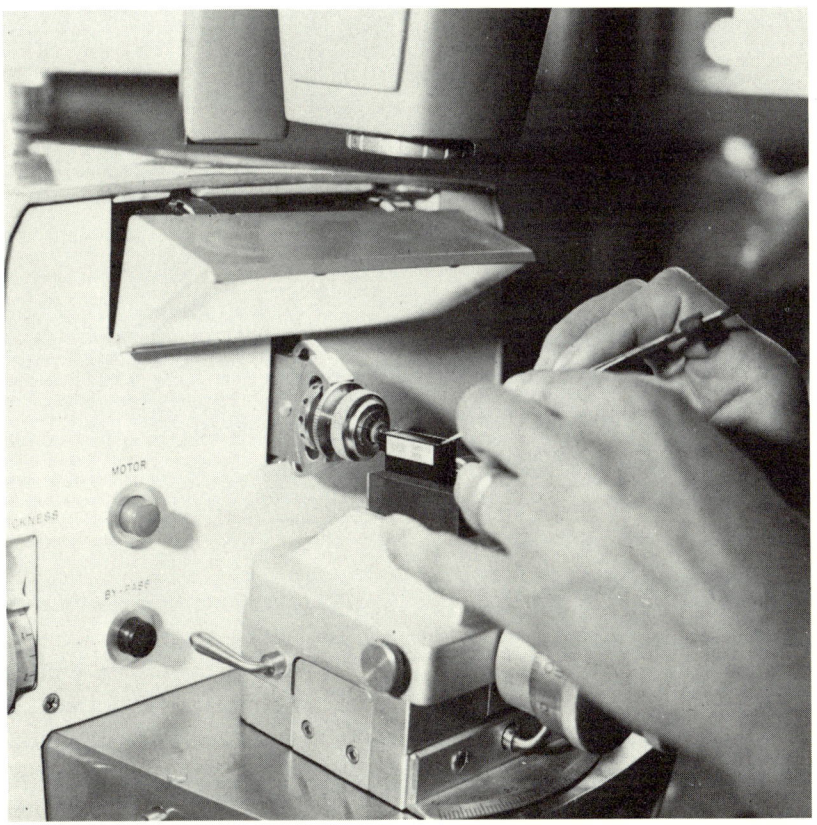

Getting the section on a grid. With the grid held by forceps, the technician dips the grid into the water, brings it up under the section, and carefully lifts the section out. The section may then be stained or shadow cast as the situation calls for. It is then put into a specimen holder as shown before.

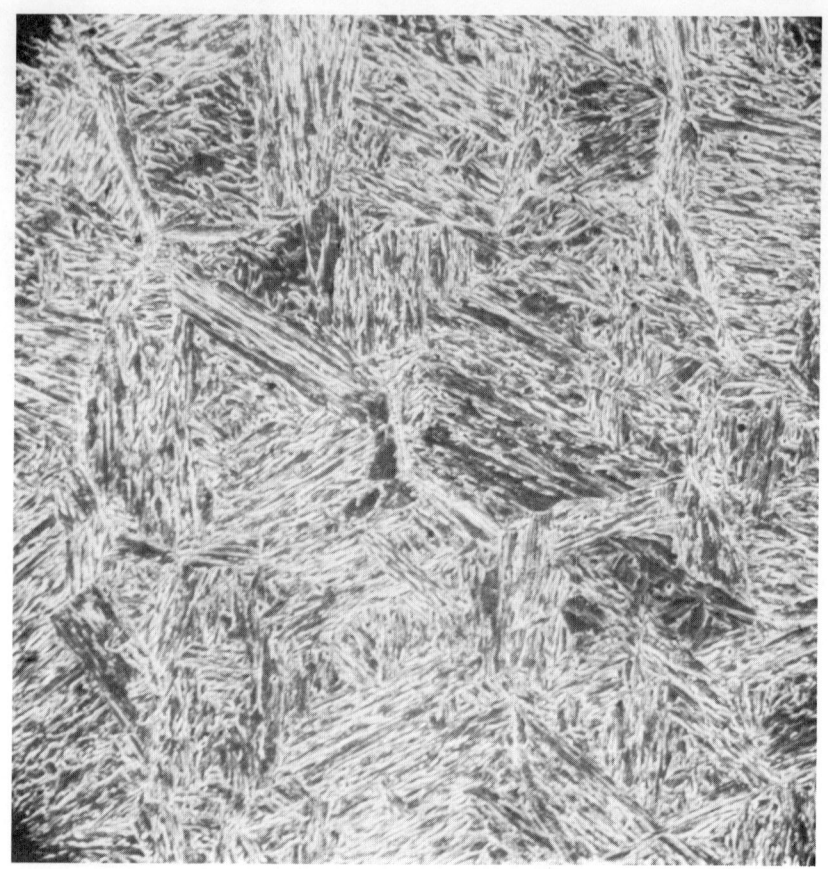
Scanning electron micrograph of polished stainless steel (900X).

Many organic materials, particularly virus particles, do not scatter electrons to any great degree. Therefore, images with good contrast are difficult to obtain. The problem is overcome with a technique called *shadowcasting*. The process is carried out with a vacuum evaporator (see page 45). A thin layer of metal is deposited on the specimen at an oblique angle. Objects, such as virus particles and bumps and ridges in the specimen, will receive a heavier deposit on the side facing the evaporator filament. The side away from the filament will be in "shadow" and will receive very little deposit of metal. When shadow-cast specimens are examined in the microscope, what is actually seen is the image of

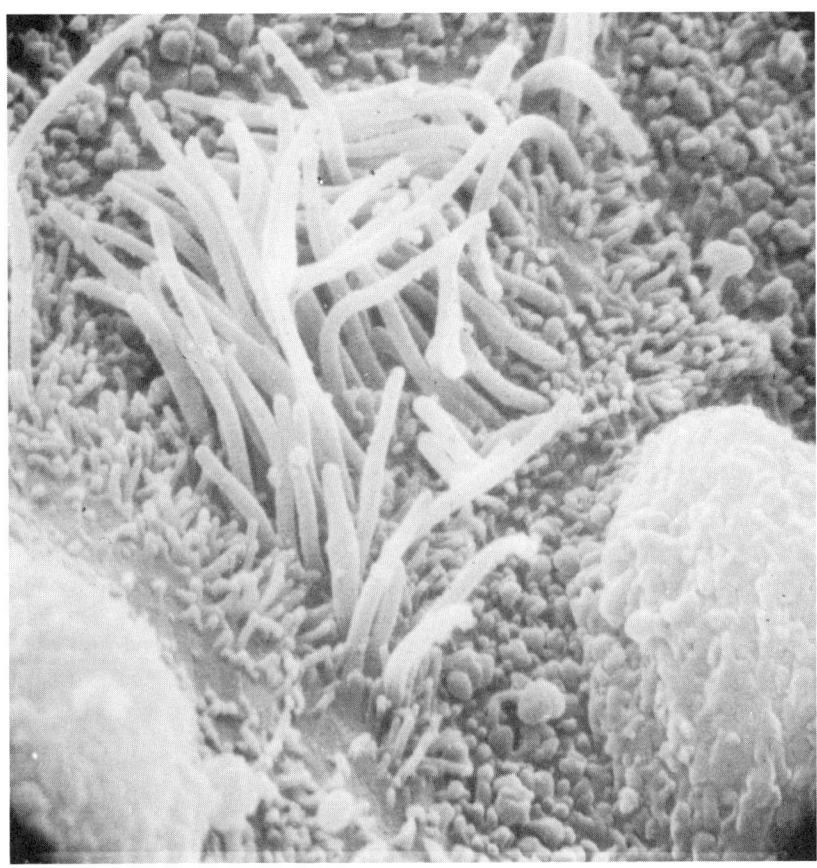

Scanning electron micrograph of surface cell from bronchial tube of a rat lung (10,000X).

the metallic deposit. The metal coating actually forms a replica of the specimen right on the grid. The areas with heavy metal deposit appear light in the image and those with no deposit appear dark. Viral particles, for example, actually appear to cast shadows. With the appropriate mathematics, the size of viruses and other objects can be calculated from the size of the "shadows."

Anyone who watches television commercials is familiar with the term "freeze-drying." Freeze-drying is a method of dehydration of materials that involves quick freezing and the removal of the water in vacuum. The water is removed by *sublimation*, that is, it passes directly from the solid to the gaseous state without an intervening

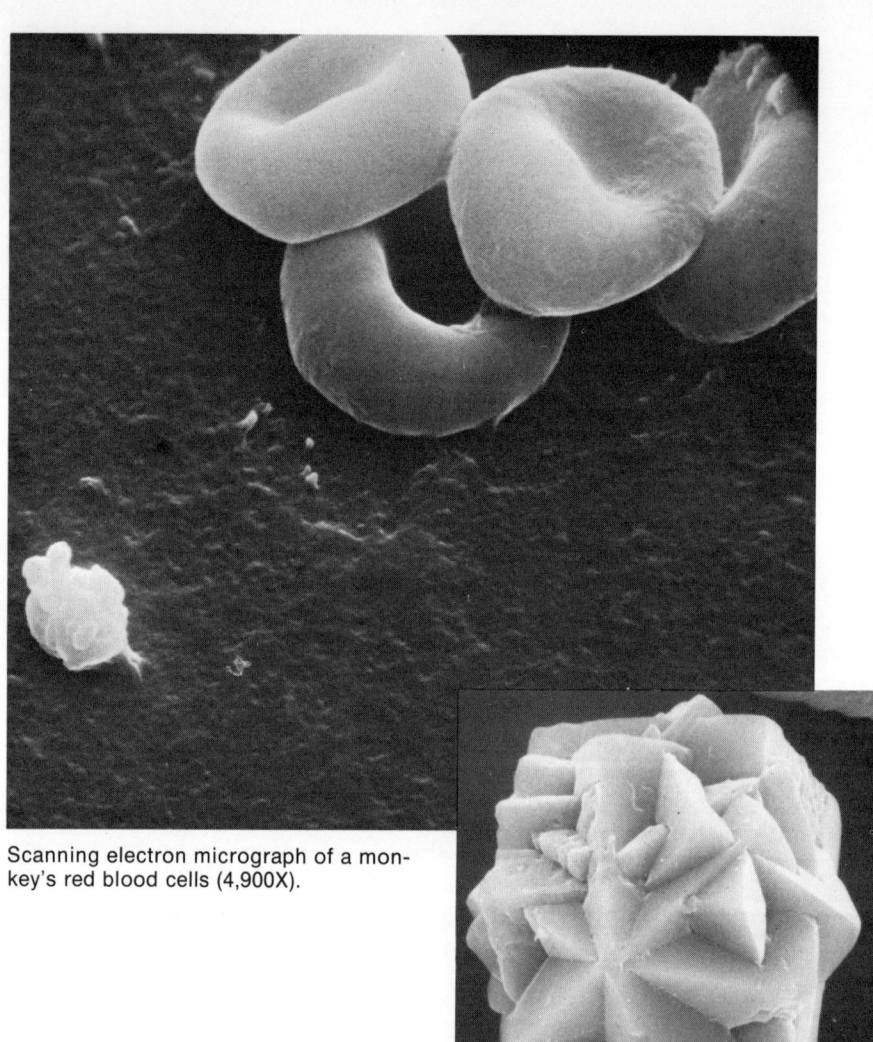

Scanning electron micrograph of a monkey's red blood cells (4,900X).

Scanning electron micrograph of zeolite crystals at 1,000X.

liquid phase. Specimens for electron microscope work usually must be free of water, and freeze-drying usually results in less distortion and deformation than most other methods. It is used for preparation of both organic and inorganic materials. One of the problems encountered in freeze-drying tissue sections is that tiny cell parts are sometimes pulled out by the "vacuum cleaner" nature of the process.

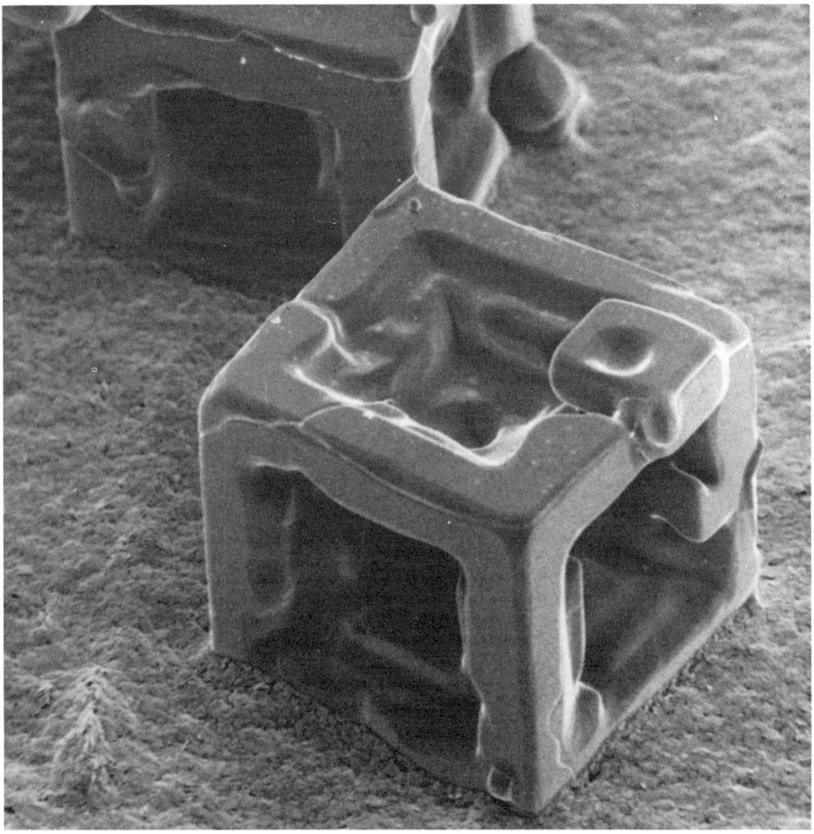

Scanning electron micrograph of potassium chloride crystals at 450X.

Scanning electron micrograph of material used for cigarette filters at 15,000X.

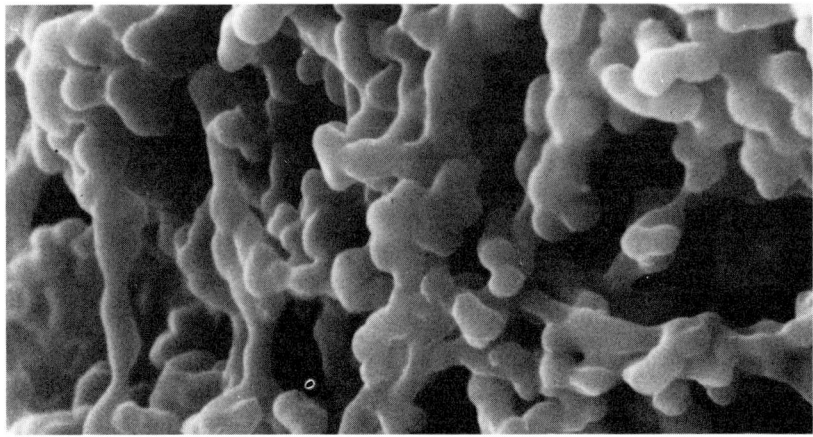

The magnifications given on electromicrographs are at best approximations. For getting an idea of how much bigger the picture is than the actual thing, the stated magnifications are adequate and have an accuracy of plus or minus 5 to 10 percent. Frequently the nature of the research requires measurement data more accurate than a 5 or 10 percent error, and electron microscope technicians are constantly seeking methods of obtaining more precise measurement of magnification.

The determination of magnification in electron microscopy is far more difficult than in light microscopy. This is part of the price paid for the increased versatility and flexibility of magnetic lenses. The focal length of magnetic lenses varies with fluctuations in the beam current and with changes in any number of the microscope's components. Even a change in the resistance of a wire as it ages can affect the instrument's performance and consequently the magnification. The focal length of optical lenses, however, remains constant no matter what the intensity of the light going through it.

One way of measuring objects in view is to compare them with something of known size. This can be done by placing the objects of known size on the grid along with the specimen. Many technicians make use of polystyrene spheres. These spheres, manufactured by chemical companies, are of remarkably constant size. Most of them are around 2400 A in diameter, give or take 500 A. Other sizes are available. The spheres are sprayed or dropped on a grid along with the specimens. The size of the sphere is compared with the size of the specimen. The preparation may also be shadow-cast and sizes can be calculated from the size of the shadows.

Many technicians dislike polystyrene spheres, complaining that they roll off the grid and contaminate the specimen chamber and other parts of the microscope. A particle 2400 A in diameter may not seem like much of a contaminent, but a collection of particles in the microscope column could cause a great deal of trouble.

Accurately ruled diffraction-gratings are also used as measurement standards. The gratings are ruled at a known number of lines per inch. Replicas are made of the gratings, mounted on grids, and used to calibrate the instrument.

There are many more techniques, and technicians are constantly developing new methods and improving on existing areas. The electron microscope technician is in great demand and is con-

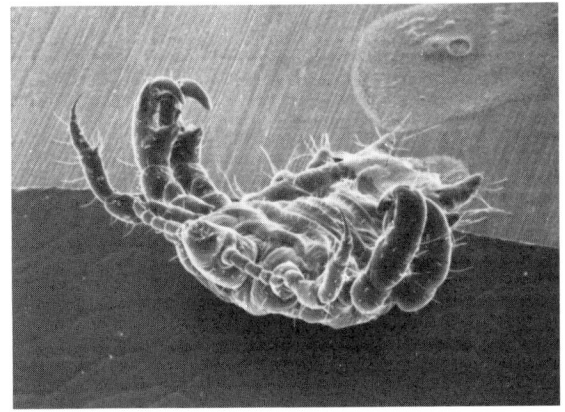

Scanning electron micrograph of human louse. *Phtirus pubis* (80X).

Scanning electron micrograph of a living aphid (60X).

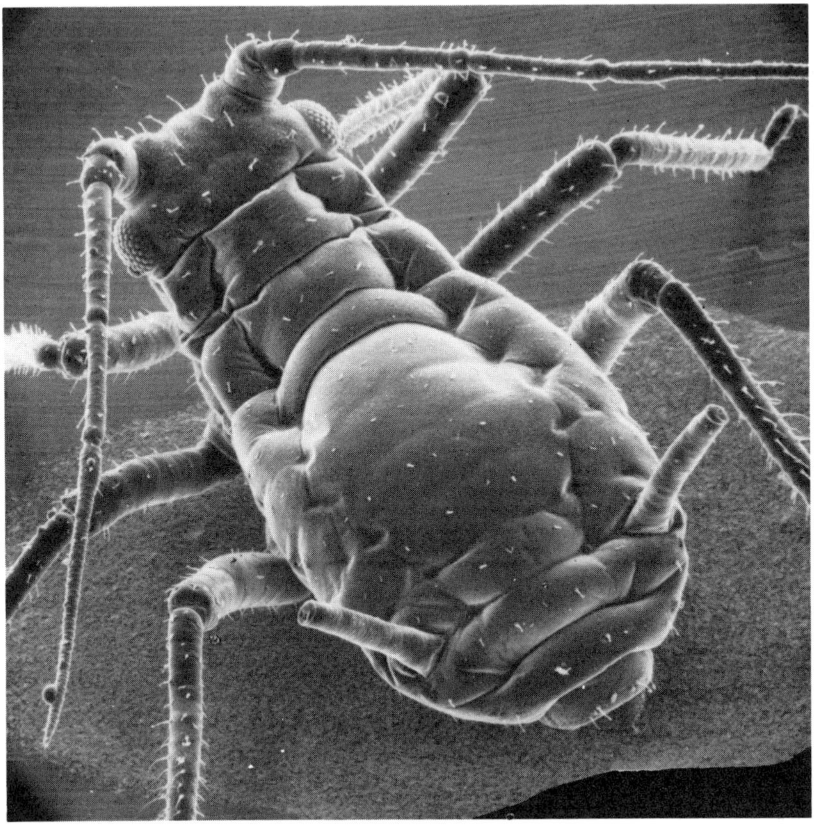

stantly challenged to devise new techniques that will enable research scientists to gain even more information with the electron microscope.

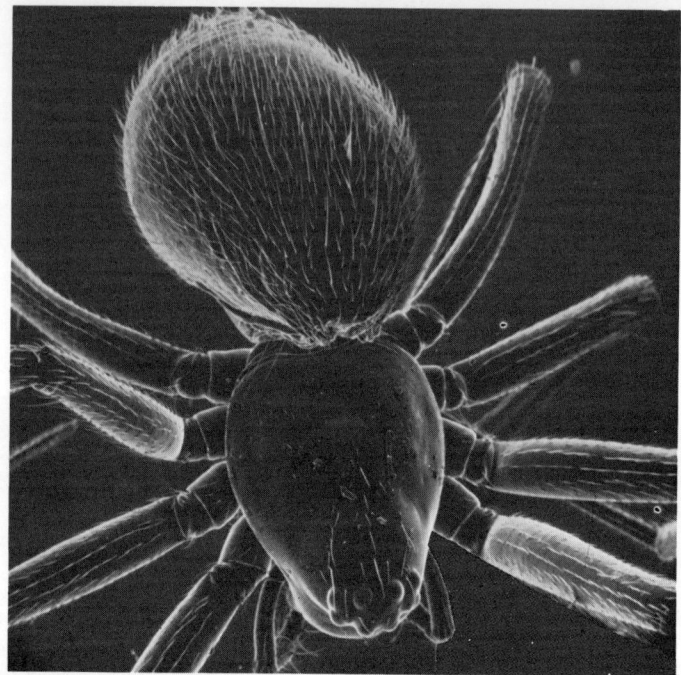

Scanning electron micrograph of a living spider (40X).

# 6
# Other Types of Electron Microscopes

While the transmission electron microscope is the type most widely used, there are many other types that yield useful information. Many of these, notably the scanning electron microscope, are increasing in popularity among research scientists in education and industry.

The emission electron microscope was developed at about the same time as transmission instruments. There are several special types of emission microscopes, used mostly by metallurgists (scientists who study metals). In the emission microscope the electrons that form the image are emitted from the specimen rather than from a filament. The specimen is placed near a heater, and it emits electrons when certain temperatures are reached. The electrons are focused by magnetic and other types of electron lenses onto the fluorescent screen or camera. The image yields information on the crystalline structures of the metal.

The field-emission microscope has actually enabled the detection of individual atoms in a very sharply pointed tungsten tip. It is essentially a sharp metal point which is the cathode. The anode in this microscope is the fluorescent screen. The electrons flow from the tip and form an image on the screen. Dotlike patterns in the image reveal the position of the individual atoms in the metal tip. The structure of organic materials such as proteins has been visualized by field emission. The technique involves vacuum evaporation of the specimen material on the tungsten tip.

The field-emission process is sometimes called "cold" emission, since it can take place at room temperature or colder. In certain situations, however, the emission is improved by heating. On the other hand, experiments in which the tip was cooled to the temperature of liquid hydrogen (-253° C) resulted in actually resolving the individual atoms in a tungsten tip. The cold kept the atoms from moving around too much, thereby yielding a sharper image and picture.

The scanning electron microscope has produced spectacular, breathtaking images. A three-dimensional effect is produced, which, besides being beautiful to look at, provides many kinds of information that cannot be supplied by the transmission electron microscope.

A great advantage of the scanning electron microscope is that the specimens need not be ultra thin. Whole objects, such as insects, can be viewed, and the preparation procedures are minimal compared to transmission electron microscopy.

A scanning electron microscope prototype was made in 1935, but these microscopes were not offered for sale until the mid-1960s. The way a scanning electron microscope works is similar in many respects to a television camera and receiver. Indeed, scanning electron microscope images are usually viewed in a television-type monitor.

The lenses of the scanning electron microscope focus the electrons into a tiny, bright spot. The spot of electrons scans the surface of the specimen in what is called a *raster* pattern. This is similar to the way the picture in a television set is scanned across the picture tube. The scanning electron microscope is sometimes called the *flying spot microscope.* The tiny spot (as small as 50 A in diameter) follows every bump, ridge, depression, and projection of the specimen surface. The image that results has great depth of field. That is, all the surface features are in focus at one time, and a three-dimensional effect is produced. As the electron spot scans the surface of the specimen, the energy of the spot "kicks out" various radiations from the specimen. These radiations include X rays, visible light, and electrons. These secondary radiations, rather than the electrons from the gun, are the major contributors to the formation of the image. Each of these secondary radiations can be used in different instruments to give a variety of information about the specimen. Visible light can give information on the chemical composition of the specimen. X rays can give information on how the atoms are arranged.

A detector device, similar in concept to a television camera, receives the radiations that issue from the specimen. The detector converts the electrons to electrical currents, which determine the various areas of light and dark. These make up the image on a cathode-ray display tube much like the picture tube of a television set. The formation of the picture is synchronized with the scanning of the specimen by the electron beam.

The scanning electron microscope is particularly useful for studying the details of the surface structure of both living and nonliving materials. For that use it is far superior to the transmission electron microscope. With the scanning scope, metal surfaces can be examined without the tedious, time-consuming process of making surface replicas.

The specimen chambers of the scanning electron microscopes are much larger than those of transmission electron microscopes. This is to accommodate the larger specimens that can be viewed in the scanning electron microscope. In many scanning electron microscopes, the specimen holder is nothing more than a metal disc. The orientation of the disc can be changed by manipulating control knobs. Although the scanning electron microscope, like any electron microscope, must be operated in vacuum, live insects have been examined. Many insects can survive a short period in a partial vacuum. Live insect specimens are glued to the specimen stage so that they will not wander off during the scanning process.

Most scanning electron microscopes can achieve magnifications up to about 100,000 diameters and resolve to about 90 A. Since the scanning electron microscope is used for large specimens, magnifications as high as those of a transmission electron microscope are not necessary. One of the advantages of the scanning electron microscope is that it can be operated at lower magnifications than transmission instruments. These lower magnifications are useful for examining miniature electronics components.

Researchers are finding more uses for scanning electron microscopes. As the demand for scanning electron microscopes increases, the number of manufacturers is increasing, too. In the 1965 *Science* Guide to Scientific Instruments, no manufacturers of scanning electron microscopes were listed. The 1972 edition of the

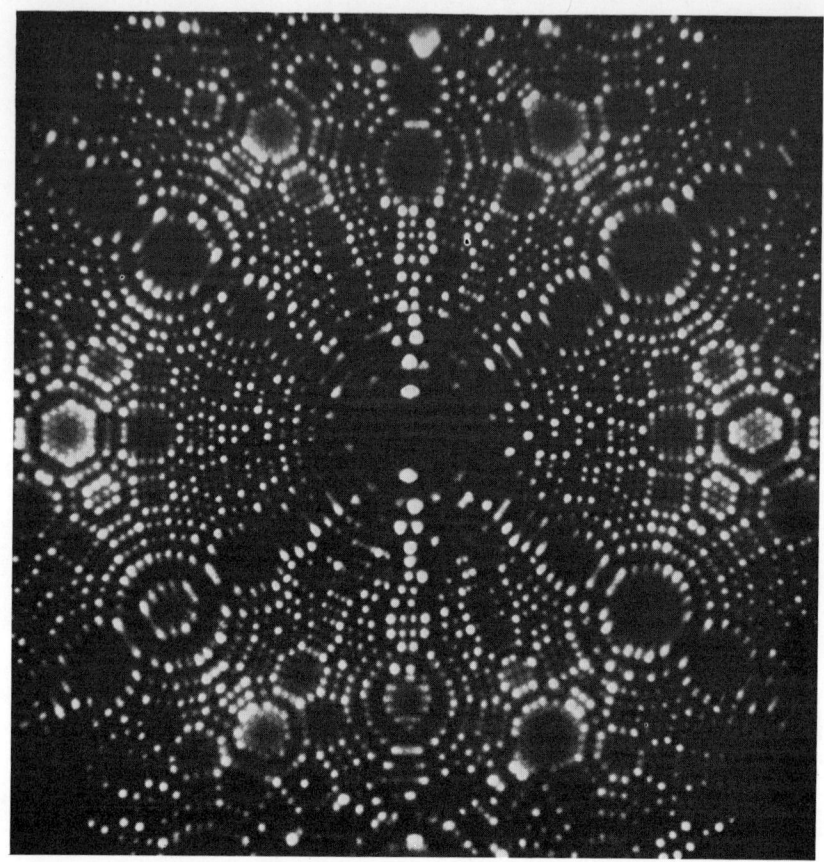

Field-ion electron micrograph showing the position of individual atoms in the tip of a tungsten filament (4,000,000X).

Guide lists over fifteen. One of the reasons for the increase is that the potential of this instrument is only being realized now.

In 1971 a scanning electron microscope, modified to provide extra high resolution, produced images of individual atoms. The instrument was built at the University of Chicago by Dr. Albert V. Crewe and his associates. With their scanning electron microscope they obtained pictures of hemoglobin molecules at a magnification of 2,750,000 diameters. Hemoglobin is the substance in red blood cells that combines with oxygen. The picture was fuzzy and the molecules had been deformed by the preparation procedure, but to biologists familiar with the structure of the molecule the blurred image represented a very exciting event. Individual

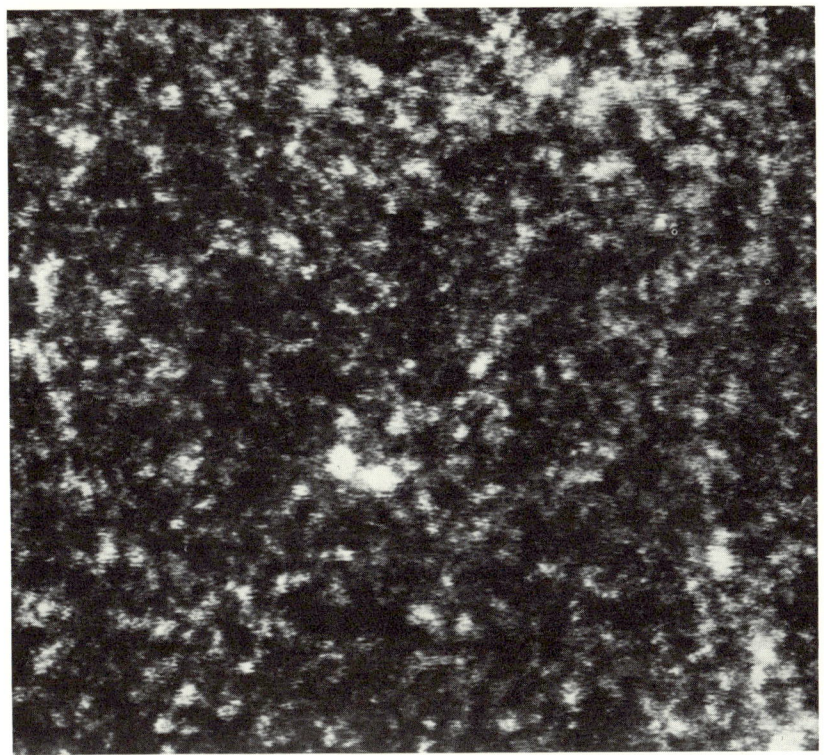

The two bright dots in the center are a pair of uranium atoms. This electron micrograph was prepared by Dr. Albert Crewe and his associates at the University of Chicago with an extremely high resolution instrument combining features of the transmission and scanning electron microscopes.

atoms of uranium were also imaged. The preparation was made by "staining" organic substances with uranyl acetate, a uranium compound. These results were obtained by an instrument which combines features of the transmission, field emission, and scanning electron microscope.

In most scanning electron microscopes resolution is not as great as in the transmission microscope. In the transmission scope the entire specimen is exposed to the electron beam for the same amount of time. In the scanning electron microscope a point of the specimen is exposed for a fraction of a second as it is scanned by the electron spot. If the resolution of scanning electron microscopes is to be improved, the electron source must be about a million times as bright as that normally obtained with the tungsten filaments usually used in the instrument.

Dr. Crewe obtained high resolutions by combining features of emission and scanning electron microscopes. He used a high-energy-field emission source and sent the electrons through an especially high vacuum. The better the vacuum, the less energy is lost by electrons striking air molecules. Since the late 1960s there have been many advances in vacuum technology. A perfect vacuum is nearly impossible to obtain. Even interstellar space is probably not a perfect vacuum. However, with the improved vacuum equipment scientists have obtained much better vacuums that are useful in a variety of research problems.

The electron source in Dr. Crewe's microscope is a tungsten wire with a diameter of about 1000 A. Electric force lines are also generated from the source and these tend to keep the electrons "on path." A 100 A spot of electrons is produced. However, 100 A is much too large to form images of atoms and most molecules. The resolution is dramatically improved with a short-focal-length magnetic lens. The builders claim a resolution of 4–6 A. The use of the lens in this way is quite similar to a transmission electron microscope.

The beam is scanned out and then transmitted through the specimen. The secondary radiations and primary electrons (from the source) are analyzed and detected by electronic devices specifically designed to get the greatest possible information from each kind of radiation.

A possible future application of instruments of this kind might be the analysis of DNA, the genetic master-molecule, in a manner so specific that individual genes could be located and identified. The vacuum and electron technology is available. New techniques of specimen preparation must be developed to take advantage of new electron microscope technology.

# 7
# Knowledge Obtained with Electron Microscopes

The light microscope revealed the "inner space" of the universe, and the electron microscope has enabled scientists to probe more deeply into this inner space.

With the light microscope biologists found out that the cell was the basic unit of structure and function of living things. They were also able to identify many cell parts and to determine the functions of those parts. With the electron microscope, biologists were able to build a picture of the minute or ultra structure of tiny parts of cells. A comparison of a cell diagram in a biology textbook of the 1940s and 1950s with a biology book of the 1960s and 1970s will quickly show how much knowledge of cell structure has been gained through the use of electron microscopes. In earlier books a cell diagram was typically a rectangle with a circle in it that represented the nucleus. Within the nucleus, a dot represented the nucleolus and various lines and shadings represented the chromatin. In the rest of the cell (cytoplasm) there were open spaces called vacuoles and various smaller parts. The smaller parts included mitochondria, chloroplasts (in plant cells) and other drawn dots and specks called *inclusions.* The term "inclusion" referred to things that could be barely seen, and whose function and structure were largely unknown.

A cell drawing from a recent biology book will show structures such as ribosomes, endoplasmic reteculum, and Golgi bodies. The structure of the mitochondria and chloroplasts are usually shown.

Strands of DNA; transmission electron micrograph at 310,000X.

*Esherichia coli*, human intestinal bacterium, with bacteriophages clustered about it; transmission electron micrograph at 33,800X.

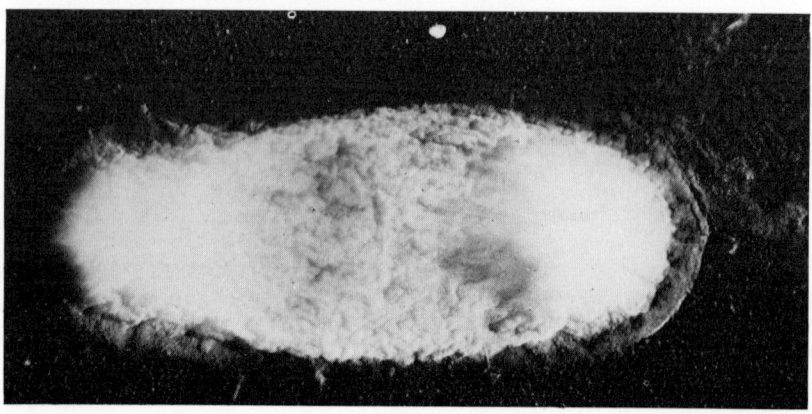

With the ultramicrotome, sections can be made so thin that mitochondria and chloroplasts are sliced through, revealing the internal structure. These cell parts are many times smaller than even the thinnest sections obtained with ordinary microtomes. Mitochondria were revealed to be sausage-shaped structures with an intricate network of double walls. Mitochondria function in the release of energy in the cell, and the twisting walls increase the surface area so that various enzymes that speed up life-sustaining chemical reactions in the cell can work more efficiently. Chloroplasts, the plant-cell structures in which the plant food-making process, photosynthesis, takes place, were shown to have an intricate "stack" structure.

In addition to clarifying the structure of cell parts the electron microscope revealed the cell as a dynamic, integrated structure. The cell membrane was shown to have a double-walled structure that directly communicates with a system of channels (endoplasmic reticulum) throughout the cell. The network of channels is continuous with the nucleus which also has a double membrane. Electron micrographs actually revealed materials leaving the nucleus and going out into the rest of the cells. These materials are nucleic-acid "messenger" molecules which carry genetic "instructions" to the cell. Tiny spherical bodies called *ribosomes* are associated with the walls of much of the endoplasmic reticulum. The synthesis, or putting together, of proteins takes place on the ribosomes. This protein synthesis is, according to theory, directed by the chemical-genetic messengers from the cell nucleus. Of course, use of the electron microscope alone did not determine the function of the various cell parts. But visualizing the cell parts did help scientists to clarify some of the details of, and to verify certain hypotheses relating to, the cell's chemical activities.

Almost everything that is known about viruses came from electron microscope work. Certainly, all knowledge of the structure of viruses was gathered by electron microscopy. Chemical analysis with radioactive tracer elements combined with electron microscopy has clarified the way in which virus particles enter cells and reproduce within cells. Biochemical analysis revealed that viruses are made of an outer part or "coat" of protein and an inner core of nucleic acid, either DNA or RNA. Work with a tadpole-shaped virus called *bacteriophage* indicated that only the nucleic acid part of the virus entered the cell. Bacteriophages are a

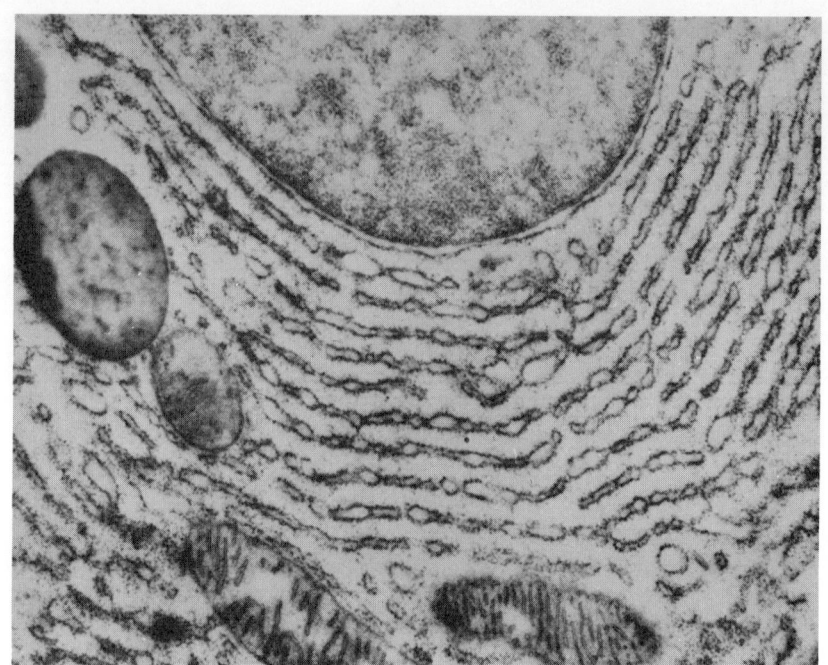
Endoplasmic reticulum in guinea pig pancreas cells (50,000X).

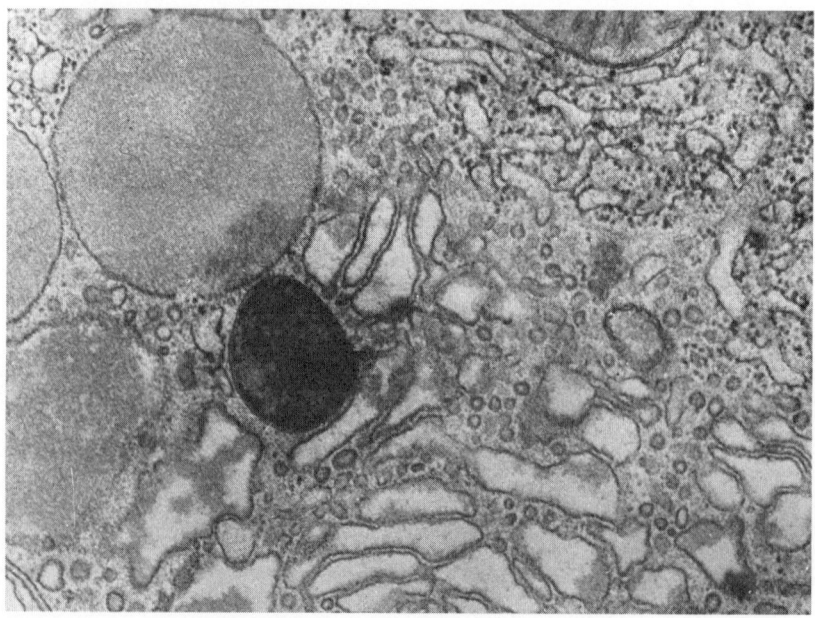
Lysozome in guinea pig pancreas cell (50,000X).

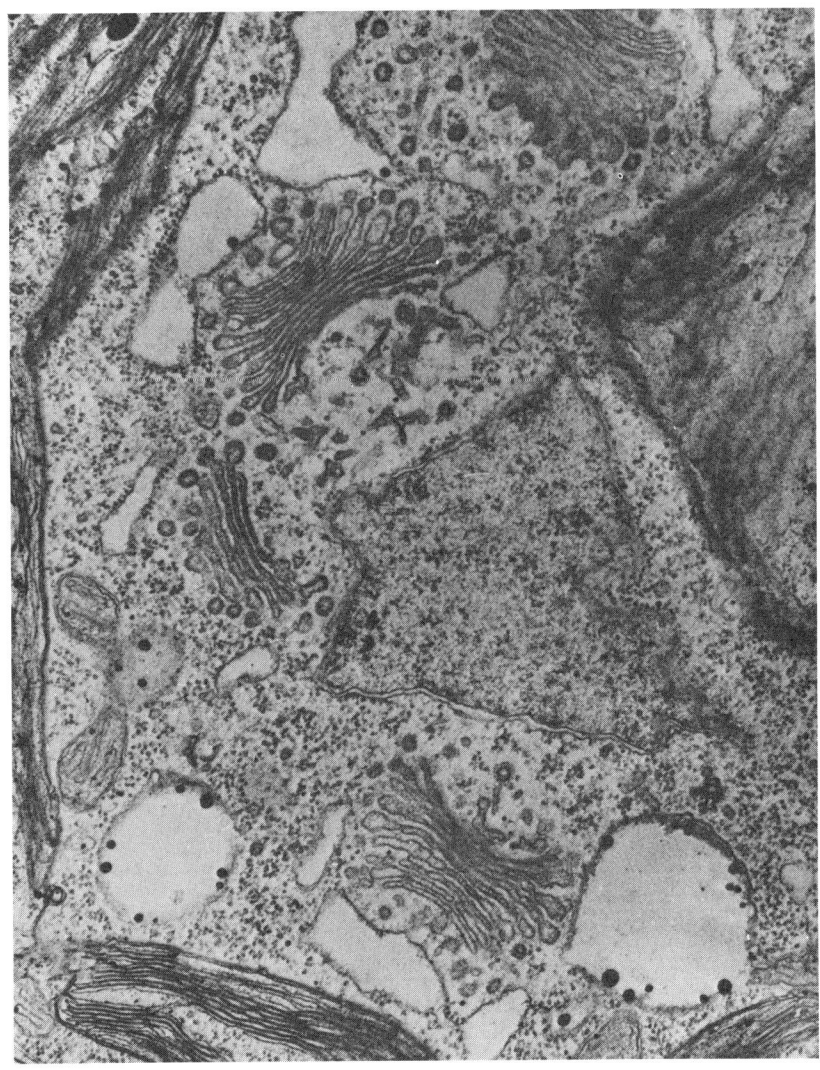

Several clusters of golgi complex can be seen in this transmission electron micrograph of guinea pig pancreas tissue (50,000X).

group of viruses that infect bacteria. These viruses attach their "tails" to the bacterial cell. The "tails" are hollow tubelike structures and the viral nucleic acid is injected into the bacteria through the tube. The protein coat remains outside the bacterium. The viral nucleic acid directs the bacterial nucleic acid to make many new viruses. In some cases the bacterial cell spectacularly bursts to release hundreds and even thousands of new virus particles.

The first molecules to be imaged and resolved by electron microscopy were large organic molecules, such as proteins. Many virus particles consist of one large protein molecule surrounding the nucleic acid. The molecules of the nucleic acids, DNA and RNA, are a favorite subject of electron microscopists. Techniques have been developed for "exploding" the entire DNA content out of a cell. Various measurements and other types of analyses can then be carried out. The possibility of locating the position of specific genes on the DNA molecule continues to excite biologists.

Physical scientists, especially metallurgists, have also made good use of electron microscopes. The development of replica techniques enabled the detection of the microstructure of metal surfaces. Metallurgists could actually see what made metals strong, weak, soft, brittle, shiny or dull. The making of alloys, mixture of different metals, became a much more exact science as metallurgists were able to see how the molecules of the metals "cling" to each other to form the alloy.

Materials such as concrete were used for hundreds of years before anyone knew what held such materials together. Knowledge gained from electron microscopy has made possible the development of more durable building materials. The mechanism of what makes glue stick surfaces together has been shown with electron microscopy.

Electron microscopes and related instruments are constantly being modified and improved to reveal even more about the inner space of the universe. The positions of individual atoms have been resolved, and with the continued evolution of these instruments, man may someday soon realize the dream of actually seeing atoms.

# Bibliography

Barer, R. and V. E. Cosslett, (ed), *Advances in Optical and Electron Microscopy.* (three volumes) New York, Academic Press, 1966, 1968, 1969.

Dawes, Clinton J., *Biological Techniques in Electron Microscopy.* New York, 1971.

Grivett, P., *Electron Optics.* New York, Pergamon Press, 1965.

Haine, M. E., *The Electron Microscope.* Interscience (John Wiley and Sons), New York, 1961.

Hall, Cecil E., *Introduction to Electron Microscopy.* Second Edition, New York, McGraw-Hill, 1966.

Heidenreick, R., *Fundamentals of Transmission Electron Microscopy.* New York, Interscience (John Wiley and Sons), 1964.

Kay, D., (ed)., *Techniques for Electron Microscopy.* Oxford, England, Blackwell Scientific Publications Ltd., 1965.

Sjostrand, F. S., *Electron Microscopy of Cells and Tissues.* New York, Academic Press, 1967.

Thornton, P. R., *Scanning Electron Microscopy.* London, Chapman and Hall Ltd., 1968.

# Index

Abbe, Ernst, 7
Accessories, for transmission-electron microscope, 37-39
Agar, 10
Angstrom, Anders Jonas, 21
Angstrom unit, 21
Astigmatism, 36

Bacteriology, microscope and development of, 10-11
Bacteriophage, 79, 81
Beam current, 32
Bell Telephone Laboratories, 25
Biology, microscope and development of, 9, 77, 79
Busch, H., 25-26, 34

Camera, electron-microscope, 37
Cathode ray, 23
Chromatic aberration, 7, 36
"Cold" emission, 72
Collodion, 45
Condenser lens, 33
Corpuscle theory of light, 19-20
Crewe, Dr. Albert V., 74, 76
Crookes, William, 23
Crookes tube, 23, 26

Davisson, Clinton, 25
de Broglie, Louis, 25, 26, 28
Decontamination device, 38
Detector device, 73
Diffraction grating measurement, 68
Direct-spraying technique, 43
Disease, 6-10
   germ theory of, 10
DNA, 79, 81-82

Einstein, Albert, 24-25
Electromagnetic lens, 34
Electromagnetic radiation, 23
Electromagnets, 26-17
Electron, 24
Electron gun, 30-32
   tungsten filament, 32
Electron microscope (transmission), 14, 29-40
   cost of, 40
   defined, 27-28
   development of, 7-8, 27
   knowledge obtained with, 77, 79, 81-82
   and optical microscope, 29
   other types of, 71-76
   scattering and vacuum, 29-30
   specimen preparation, 28, 30
   structure of, 30
   techniques, 41-53, 64-70
   voltage, 30-31, 32-33
Elford, William, 12
Embedding process, 51
Emission electron microscope, 71

Field-emission microscope, 71-72
Filterable virus, 12
Flying spot microscope, 72
Freeze-drying technique, 65-66

Germ theory of disease, 10
Germer, Lester, 25
Golden Age of Bacteriology, 10
Goniometer, 38
Grid, 43
Grid spade 44-45

Hillier, James, 27
Histology, 41

Hooke, Robert, 4, 5, 14
Huygens, Christian, 20

Inclusions, 77, 79
Instruments, scientific, importance of, 1

Janssen, Hans, 3
Janssen, Zacharias, 3

Knives, 50-51
Knoll, R., 26-27, 34
Koch, Robert, 10-12

Leeuwenhoek, Anton van, 3, 4, 5, 9, 18
Lenses, 18-19
Light, nature and theories of, 15-22, 24, 25
Light microscope, 3
   histological techniques, 41

Magnetic electron lenses, 26-27
Magnification, and determination of, 68
Metallurgy, 82
Microbes, 10, 12
*Micrographia,* 5
Micron, defined, 21
Microscope, light, 3
   compound, description of, 5-6
   phase-contrast, 7
Microtomes, 41, 42, 49-50
   knives, 51
Millimeter, definitions, 21

Neutron, 24
Newton, Isaac, 15, 19, 25

Objective lens, 34
Oil immersion objective, 7

Pasteur, Louis, 10
Phase-contrast microscope, 7
Photoelectric effect, 24-25
Photon, 24
Planck, Max, 24
Polystyrene spheres, 68
Prebas, A., 27
Prism, 15
Projector lens, 36-37
Proton, 24

Quanta, 24
Quantum theory, 24-25

Radio Corporation of America, 27
Raster pattern, 72
Refraction, 16, 18
Replicas, 43, 47-48
Resolution (resolving power), 18-19, 22
   innovations in, 74-76
Ribosomes, 79
RNA, 79, 81-82
Rudenberg, 27
Ruska, Ernst, 26-27, 34

Scanning electron microscope, 30, 72
   advantages of, 72
   function, 72-73
   magnification, 73
   resolution innovations, 74-76
   specimen chamber, 73
   and vacuum technology, 76
Science, 23-24
   electron microscope and, 13-14
Seneca, 18
Shadowcasting, 64-65
Siemens Company, 27
Specimen
   chamber, 33
   holders, 33-34, 38
   preparation techniques, 47
   size determination, 68
Spherical aberration, 34
Spread-film technique, 44-45
Stains, electron microscope, 47
Stanley, Wendell, 12-13
Stereo viewing, 37
Stigmator, 36
Sublimation, 65

Techniques, electron microscope, 41-53, 64-70
Television monitoring, 37
Thin sectioning, 43, 49
Thomson, Joseph John, 24, 26
Transmission electron microscope
   *See* Electron microscope

Ultramicrotomes, 50, 79

Vacuum
   evaporator and evaporation, 45

Vacuum:
  technology, 76
Video taping, 37
Viewing screen 37
Viruses, 12-13, 28, 79, 81

Wavelength, 21-22

Wave theory of light, 20
Wehnelt cylinder, 32
Wet specimen viewing, 38-39

Young, Thomas, 20-21

Seneca East Jr. High School

502.8
KLE

Klein, Aaron E.

The electron mi-
croscope

DATE DUE